AF522301

GENETIC ENGINEERING OF PLANTS

GENETIC ENGINEERING OF
PLANTS

GYAN DEEP SINGH

ANMOL PUBLICATIONS PVT. LTD.
NEW DELHI - 110 002 (INDIA)

ANMOL PUBLICATIONS PVT. LTD.
H.O.: 4374/4B, Ansari Road, Daryaganj,
New Delhi-110 002 (India)
Ph.: 23278000, 23261597
B.O.: No. 1015, Ist Main Road, BSK IIIrd Stage
IIIrd Phase, IIIrd Block,
Bangalore - 560 085 (India)
Visit us at: www.anmolpublications.com

Genetic Engineering of Plants

ISBN 978-81-261-3599-8

PRINTED IN INDIA

Printed at Mehra Offset Press, Delhi.

Contents

Preface

This book is designed as an introductory text in botany. It assumes little knowledge of the sciences on the part of the student. It includes sufficient information for some shorter introductory botany courses open to both majors and nonmajors, but it is arranged so that certain sections-for example, "Soils," "Molecular Genetics," "Division Psilotophyta. Botany instructors vary greatly in their opinions concerning the depth of coverage needed for the topics of photosynthesis and respiration in a text of this type. Some feel that nonmajors, in particular, should have a brief introduction only, while others consider a more detailed discussion essential. In this text, photosynthesis and respiration are discussed. Despite eye-catching chapter titles and headings, many texts for majors and nonmajors give relatively minor coverage of the current interests of a significant number of students.

This text emphasizes current interests without giving short shrift to botanical principles. Present interests of students include subjects such as global warming, ozone layer depletion, acid rain (acid deposition), genetic engineering, organic gardening, Native American and pioneer uses of plants, pollution and recycling, house plants, backyard vegetable gardens, natural dye plants, poisonous and hallucinogenic plants, and the nutritional values of edible plants. The rather perfunctory coverage or absence of such topics in many botany texts has occurred partly because botanists previously have tended to believe that some of the topics are more appropriately covered in anthropology and horticulture courses. I have found, however, that both majors and nonmajors in botany, who may be initially disinterested in the subject matter of a required course, frequently become engrossed if the material is repeated become engrossed if the material is repeatedly related to such topics. Accordingly, a considerable amount of ecological and ethnobotanical materials has been included with traditional botany throughout the book-without, however, resorting to excessive use of technical terms.

Author

Chapter 1

Introduction and Nutrition of Plants

During the last 30 years, production of the main food crops has doubled. This increase of production has mainly been achieved by introduction of high-yielding varieties, irrigation, and the use of fertilizers and pesticides. Due to this increase, the share of people in developing countries with insufficient average food supply has decreased from 74 percent in 1962 to 6 percent in 1988, representing 230 million people. In many regions of the world, the intensification of crop production has led to deterioration of soil fertility, erosion, salinization, reduction of biodiversity, and other deleterious side-effects. The use of pesticides has more than tripled since 1970 and is a growing concern especially in developed countries. Despite the intensive use of chemical crop protection methods, the losses due to pests, pathogens and weeds are more than 40 percent of attainable production, representing a value of more than 240 billion US$.

Genetic engineering offers new possibilities for the breeding of plant varieties with increased resistance to pests and pathogens. New resistant varieties may lessen the dependence on pesticides and help securing sufficient crop yields in the future.

In plant genetic engineering, genes from different organisms (other plants, bacteria, viruses, etc.) are transferred into the genome of a plant cell. The bacterium Agrobacterium tumefaciens is frequently used as a vehicle for the introduction of foreign DNA into the plant genome. In nature, these bacteria transfer some of their genes into the plant genome, thereby inducing a plant disease which leads to production of compounds used by the bacteria. In genetic engineering, the genes causing disease are replaced by genes conferring other traits. For some plants, e.g. wheat or maize, Agrobacterium-mediated gene transfer is difficult or not possible. In these cases, a technique called particle bombardment is often applied. In this method, gold or tungsten particles of about 5 μm in diameter are coated with DNA and shot into plant cells, where the DNA is released and incorporated into the plant genome. After incorporation of the foreign gene, a plant is regenerated from an engineered cell, and the traits coded for by the transferred gene are expressed by the plant.

Resistance of transgenic plants to insect pests or diseases has been achieved in more than 20 different crops, including maize, potato, squash, cotton, soybean, oilseed rape, tomato, tobacco, alfalfa, rice, barley and others. Very high levels of resistance to insect pests and viral diseases have been reached.

Insect resistance has mostly been obtained by using a gene derived from the common soil bacterium Bacillus thuringiensis. This bacterium produces a protein called Bt toxin which is toxic for certain insects. Intensive investigations have led to a detailed knowledge of the mechanism and specificity of toxin activity. In several studies, no effect of Bt toxin on humans, other mammals, and most non-target insects could be shown. Transgenic plants expressing Bt toxin were found to be protected against repeated heavy infestations of the target insect pest which totally devastated non-transgenic control plants. Other approaches to insect resistance focus on the use of genes which are part of the natural defence system of plants. The products of these genes interfere with insect digestion. Plant-derived protease inhibitors prevent protein degradation, and amylase inhibitors block starch- degrading enzymes in the insect midgut. Some of these strategies have proven to be effective and may soon be used in the development of commercial varieties.

Virus resistance is mostly achieved by introducing gene sequences derived from pathogenic viruses into the crop genome. The introduction of genes coding for viral coat proteins has been very successful. During the last years, this strategy has led to a number of crop varieties resistant to important plant viruses. More recently, also other viral genes were found to confer resistance, e.g. replicase genes, defective viral genes or antisense coat protein genes. The mechanisms of resistance are not yet completely understood.

Strategies applied to achieve fungal resistance make use of plant genes acting on different levels of the plant defence system against pathogens. Several of these strategies have led to increased resistance, but so far the level of protection was mostly to low to be of agronomic importance. Chitinase and glucanase genes coding for enzymes which break down fungal cell walls have been used in several crops including rice and have led to significant protection in some cases. The growing understanding of plant defence mechanisms is expected to lead to increased levels of protection in the near future.

Also methods investigated to obtain resistance to bacteria have not led to high levels of protection yet. Reduction of disease development in tobacco was achieved by transferring a cecropin gene derived from the Giant silk moth. Cecropins are produced by insects to fight pathogen attack and had a similar effect in some plants. Other partially successful strategies make use of genes which code for toxindetoxifying enzymes or plant genes involved in the response to pathogen attack.

Besides genetically engineered plants, also viruses and bacteria have been genetically altered in order to develop new crop protection methods. Baculoviruses are insect pathogens which have been used as a biological pesticide since the 1930s. As these viruses may take weeks to kill their host after infection, their usefulness has been limited. By transferring genes coding for insect-specific toxins, insect hormones and insect enzymes into the virus genome, the killing time has been reduced by up to 50 percent, which is not enough to achieve sufficient protection. Bacillus thuringiensis (Bt) toxin genes have been introduced into different bacteria for Bt toxin delivery to insect pests. In one approach, transgenic bacteria expressing Bt toxin are killed and then sprayed on the crop plants like a pesticide. Another approach uses bacteria living inside of plants for Bt toxin delivery.

The safety aspects of transgenic organisms have been discussed and investigated since the first successful gene transfer in the early 1970s. The release of transgenic plants is subject to different legal regulations. Before a transgenic crop may be released, potential hazards like the possibility of gene transfer to other plants or microorganisms, weediness of the engineered crop, and the expression of undesirable traits resulting from secondary effects of the gene insertion are examined. Also possible toxic and allergenic effects are analyzed, especially if the engineered plant is destined to serve as a food crop. So far, no deleterious effects of transgenic plants or other organisms have been reported.

Between 1986 and 1993, more than 1000 field releases of transgenic plants were conducted in 32 countries, and the number is increasing rapidly. The USA and Canada account for more than half of all trials recorded. Thirty-eight different plant species have been tested. Potato, oilseed rape, tobacco, maize, tomato and sugar beet constitute together 70 percent of all trials. Among these crops, about one third of all trials evaluated pest and disease resistance (16 percent virus resistance, 13 percent insect resistance, 3 percent fungal resistance and 1 percent bacterial resistance). The trait most commonly tested was herbicide tolerance (34 percent). Other traits examined were quality improvement (20 percent) and marker genes (10 percent).

In the USA, permits for commercialization of 10 transgenic crop varieties have been granted as of July 1995: Insect resistant potato, maize and cotton expressing Bt toxins, virus resistant squash expressing viral coat proteins, two tomato varieties with extended shelf-life, one tomato variety with enhanced process value, oilseed rape with altered oil composition, and herbicide tolerant cotton and soybean. In Canada, herbicide tolerant flax and oilseed rape have been commercialized. In the European Union, a herbicide tolerant tobacco variety has obtained a permit for market introduction.

Several other transgenic crops are approaching commercialization. In the field of pest and disease resistance, it is likely that more insect resistant crops expressing Bt toxins or virus resistant crops engineered with viral genes will enter the market in the near future. Within some years, varieties with enhanced resistance against fungal and bacterial pathogens may also become available. Other applications of transgenic plants which may reach the marketplace within some years include e.g. cotton with altered fibers, crops with improved nutritional value and plants producing biodegradable plastic, cheap vaccines and pharmaceuticals.

PLANT NUTRITION

Plant nutrition and the soil-plant system. The key-role of fertilizers and their judicious use in crop husbandry is well understood, when one is familiar with the general facts about plant nutrition. It is now known that at least 16 plant-food elements are necessary for the growth of green plants. These plant-nutrients are called essential elements. In the absence of any one of these essential elements, a plant fails to complete its life cycle, though the disorder caused can, however, be corrected by the addition of that element.These 16 elements are: Carbon(C), hydrogen(H), oxygen(O), nitrogen(N), phosphorous(P), sulphur(S), potassium(K), calsium(Ca), magnesium(Mg), iron(Fe), manganese(Mn), zinc(Zn), copper(Cu), molybdenum(Mb), boron(B) and chlorine(Cl). Green plants obtain carbon from carbon-di-oxide from the air; oxygen and hydrogen from water, whereas the remaining elements are taken from the soil. Based on their relative amounts, normally found in plants, the plant nutrients are termed as macronutrients, if large amounts are involved, and micronutrients, if only traces are involved. The micronutrients essential for plant growth are iron, manganese, copper, zinc, boron, molybdenum, and chlorine. All other essential elements listed above are macronutrients.

As mentioned above, most of the plant nutrients, besides carbon, hydrogen and oxygen, originate from the soil. The soil system is viewed by the soil scientists as a triple-phased system of solid, liquid and a gaseous phases. These phases are physically seperable. The plant nutrients are based in the solid phase and their usual pathway to the plant system is through the surrounding liquid phase, the soil solution and then to the plant root and plant cells. This pathway may be written in the form of an equation as: M(Solid)->M(Solution)->N(Plant root)->(Plant top) where 'M' is the plant nutrient element in continual movement through the soil-plant system. The operation of the above system is dependent on the solar energy through photosynthesis and metabolic activities. This is however, an oversimplified statement for gaining a physical concept of the natural phenomenon, but one should bear in mind that there are many physico and physico-chemical processes influencing the reactions in the pathway. The actual transfer in nature takes place through the charged ions, the usual form in which plant-food elements occur in solutions(liquid phase of the system). Plant roots take up plant-food elements elements from the soil in these ionic forms. The positively charged ions are called 'cations' which include potassium(K^+), Calcium(Ca^{++}), magnesium(Mg^{++}), iron(Fe^{+++}), zinc(Zn^{++}), and so on. The negatively charged ions are called anions and the important plant nutrients taken in this form include nitrogen(NO^-_3), phosphorous($H_2PO^-_4$), sulphur(SO^-_4), Chlorine(Cl), etc.

The process of nutrient uptake by plants refers to the transfer of the nutrient ions across the soil root interfaces into the plant cell. The energy for the process is provided by the metabolic activity of the plant and in its absence no absorption of nutrients take place. Nutrient absorption involves the phenomenon of ion exchange. The root surface, like soil, carries a negative charge and exhibits cation-exchange property. The most efficient absorption of the plant nutrients takes place on the younger tissues of the roots, capable of growth and elongation.

In this respect, root-systems are known to vary from crop to crop. Hence their feeding power differs. The extent and the spread of the effective root-system determines the soil volume trapped in the feeding-zone of the crop plant. This is indeed an important information in a given soil-plant system which helps us to choose fertilizers and fertilizer-use practices. The absorption mechanisms of the crop plants are fairly known now. There are three mechanisms in operation in the soil-water-plant systems.

They are:

1. The contact exchange and root interception,
2. The mass flow or convection, and
3. *Diffusion*: In the case of contact exchange and root interception, the exchangeable nutrients ions from the clay-humus colloids migrates directly to the root surface through contact exchange when plant roots come into contact with the soil solids.

Nutrient absorption through this mechanism is, however, insignificant as most of the plant nutrients occur in the soil solutions. Scientists have found that plant roots actually grow to come into contact with only 3 percent of the soil volume exploited by the root mass, and the nutrient uptake through root interception is even still less. The second mechanism is mass flow or convection, which is considered to be the important mode of nutrient uptake. This mechanism relates to nutrient mobility with the movement of soil water towards the root surface where absorption through the roots takes place along with water. Some are called mobile nutrients. Others which move only a few millimetres are called immobile nutrients. Nutrient ions such as nitrate, chloride and sulphate, are not absorbed by the soil colloids and are mainly in solution. Such nutrient ions are absorbed by the roots along with soil water. The nutrient uptake through this mechanism is directly related to the amount of water used by the plants (transpiration). It may, however, be mentioned that the exchangeable nutrient cations and anions other than nitrate, chloride and sulphate, which are absorbed on soil colloids are in equilibrium with the soil solution do not move freely with water when it is absorbed by the plant roots. These considerations, therefore, bring out that there are large differences in the transport and root absorption of various ion through the mechanism of mass flow. Mass flow is, however, responsible for supplying the root with much of the plant needs for nitrogen, calcium and magnesium, when present in high concentrations in the soil solution, but does not do so in the case of phosphorous or potassium. The nutrient uptake through mass flow is largely dependent on the moisture status of the soil and is highly influenced by the soil physical properties controlling the movement of soil water.

The third mechanism is diffusion. It is an important phenomenon by which ions in the soil medium move from a point of higher concentration to a point of lower concentration. in other words, the mechanism enables the movement of the nutrients ion without the movement of water. The amount of nutrient-ion movement in this case is dependent on the ion-concentration gradient and transport pathways which, in turn, are highly influenced by the content of soil water. This mechanism is predomionant in supplying most of the phosphorous and potassium to plant roots. It is important to note that the rhizophere volume of soil in the immidiate neighbourhood of the effective plant root receives plant nutrients continously to be delivered to the roots by diffusion. However, when the nutrient concentration builds up far excess of the plant in the reverse direction. These are some of the choice of fertilizers and fertilizer practices for practising scientific agriculture.

The relationship in the soil-plant system stated in the simple equation give in the earlier paragraph reflects the highly dynamic nature of the soil solution. One knows that the roots of the growing plants continuously remove nutrient ions from the soil solutions. At the same time, the breakdown of the soil minerals and the generating of more exchangeable cations, the biological activity and the additions made to the anions, e.g. nitrates, continuously change the composition of the soil solution. At a given point of time, therefore, the available plant nutrients in the soil solution may range from a tiny amount to larger quantities. Under favourable conditions, crop plants, in general, require larger amounts of plant nutrients than the quality found in soil solution at any given time. Hence, the situation of nutrients supply to plants becomes a limiting factor, specially, at the critical stages of plant growth and low crop yeilds result in recognition, therefore, fertilizers application and the use of suitable fertilizers are recommended for higher crop yeilds in productive farming. The knowledge of the specific role of each essential element in the growth of crop plants and their amounts required for efficient crop production is considered necessary in adopting scientific fertilizers use.

PLANT NUTRIENTS AND THEIR FUNCTIONS

The plants require, the following essential nutrients for their normal development:

Carbon	Nitrogen	Calcium
Hydrogen	Phosphorous	Magnesium
Oxygen	Potassium	Sulphur
Iron	Zinc	Chlorine
Manganese	Boron	..
Copper	Molybdenum	..

Carbon is obtained from carbon dioxide of the air; Oxygen from air and water; Hydrogen from water; Nitrogen from air and soil or both, and all other nutrients from the soil. Soil is a the most important source of plant food.

Nitrogen, Phosphorous and Potassium are known as primary plant nutrients; Calcium, Magnesium and Sulphur are secondary nutrients; Iron, Manganes, Copper, Zinc, Boron, Molybdenum and Chlorine as trace elements or micronutrients. The primary nutrients and secondary nutrients elements are known as major elements. This classification is based on their relative abundance, and not their relative importance. the micronutrients are required in small quantities, but they are as important as the major elements in plant nutrients.

Air is the primary source of Nitrogen for plant nutrient. Only leguminous crops can directly use this free Nitrogen with the help of symbiotic bacteria of the genus *Rhizobium*. Other plant derive from soil their Nitrogen in the form of Nitrogen and Ammonium. Nitrogen and Ammonium are produced in the soil by action of micro-organisms on the soil organic matter. Non symbiotic micro-organisms can fix free Nitrogen of the air and make it available to plant in Ammonium and Nitrates forms.

Nitrogen: Nitrogen encourages the vegetative development of plants by importing a healthy green colour to the leaves. It also controls, to some extent the efficient utilization of phosphorous and Potassium. Its dependency retards growth and root development, turns the foilage yellowish or pale green, histens maturity, causes the shrivelling of grains and lowers crop yield. The older leaves are affected first. An excess of Nitrogen produces leathery (sometimes crinkled), dark-green leaves and succulent growth. It also delays the maturation of plants, impairs the quality of crops like barley, potato, tobacco, sugarcane, and fruits; increases susceptibility to diseases and causes 'lodging' of cereal crops by inducing an undue lengthoning of the stem internodes.

Phosphorous: Phosphorous influences the vigour of plants and improves the quality of crops. It encourages the formation of new cells, promotes root growth(particularly the development of fibrous roots), and hastens leaf development through emergence of ears, the formation of grains, and the maturation of crops. It also increases resistence to diseases and strengthens the stems of cereal plants, thus reducing their tendency to lodge. It offsets the harmful effects of excess nitrogen in the plant. When applied to leguminous crops, it hastens and encourages the development of nitrogen-fixing nodule bacteria. If phosphorous is deficient in the soil, plants fail to make a quick start, do not develop a satisfactory root-system, remain stunted and sometimes develop a tendency to show a reddish or purplish discolouration of the stem and foilage owing to an abnormal increase in the sugar content and the formation of anthoscyanin.

However, the deficiency of this element is not easily recognised as that of nitrogen. It has also been observed that cattle feeding on the produce of deficient soils become dwarfed, develop stiff joints and lose the velvetty feel of the skin. Such animals show an abnormal craving for eating bones and even soil itself.

Potassium: Potassium enhances the ability of plants to resist diseases, insect attacks, and cold and other adverse conditions. It plays an essential part in the formation of starch and in the production and translocation of sugars, and is thus of special value to carbohydrate-rich crops, e.g. sugarcane, potato and sugar-beet. The increased production of starch and sugar in legumes fertilized with potash benefits the symbiotic bacteria and thus enhances the fixation of nitrogen. It also improves the quality of tobacco, citrus etc. With an adequate supply of potash, cereals produce plump grains and strong straws. But an excess of element tends to delay maturity, though, not to be the same extent as nitrogen.

Plants can make up and store potassium in much larger for correcting zinc deficiency. The symptoms of zinc deficiency appear generally in younger leaves, starting with intervienal chlorosis leading to a reduction in shoot growth and the shortening of internodes. Mottle leaf, little leaf, etc. in the case of trees are symptoms of zinc difficiency. The buds of several defficient maize plants become white; in citrous interveinal chlorosis and mottled leaf occur. In calcareous soils and in soils with very high phosphorus content, zinc defficiency is commonly expected to occur. The principal function of zinc in plants is as a metal activator of enzymes. In highly weathered coarse textured soils zinc deficiency appears under an intensive cropping programme. The availability of zinc is least between pH 5.5 and 7, but its availability increases at a lower pH. At a higher pH above 7, zinc availability becomes a complex problem, as the positively charged zinc ion gets converted into a negatively charged zincate complex whose availability tends to be reduced in alkaline soils. When the calcium ion is predominent, the highly insoluble calcium zincate is formed and zinc availability gets seriously limited. The application of soluble zinc salts or zinc chelates to the soil is generally recommended to correct its defficiency. Foliar sprays are advocated, especially for orchard trees for amending zinc defficiency. About 5 to 50 kg of zinc sulphate per hectare is used for such purposes.

The symptoms of *boron* defficiency vary with the kind and age of the plant, the conditions of growth and the severity of the deffiency. Each crop produces its characteristic growth abnormalities associated with boron defficiency, such as yellows and rosetting in lucerne, snakehead in wallnuts, die-back and corking of fruits in apple, corking and pitting of fruits in tomatoes, hollow stem and the bronzing of curd in cauliflower, the brown-heart diseases in table-beets, turnips, etc.

Molybdenum deficiency produces whip-tail in cauliflower, broccoli and other *Brassica* crops. The deficiency of this element reduces the activity of the symbiotic and non-symbiotic nitrogen-fixing micro-organisms.

It was in 1954 that *chlorine* was proved to be an essential micronutrient. Its defficiency under field conditions has not been reported so far. In water-culture solutions, the leaves of chlorosis, necrosis and an unusual bronze discolouration on tomatoes.

Sodium is not an essential element for plant growth. But some crops, such as beet, celery, cabbage, kale, knol-khol, radish, rape and turnip, benefit greatly by application of soluble sodium salts, specially if the soil is deficient in potassium. Sodium is also of direct benefit to plants indigneous to the sea-shore or to irrigated arid regions. Salts of this element are said to release more of potassium from the exchange complex and to help to maintain phosphorus in a more available form. They also serve as a partial substitute for potassium in the case of potatoes and cotton.

MAINTENANCE OF SOIL FERTILITY

No two soils are alike either in respect of their nature or in respect of quantities of plant nutrients they contain. Under a given situation, the system of farming, soil management and manuring practices, etc., influence the productiovity of soils and crop yields obtained from them.

It is estimated that the different agricultural crops in India remove about 4.27 million tonnes of nitrogen, 2.13 million tonnes of phosphoric acid, 7.42 million tonnes of potash and 4.88 million tonnes of lime per year. The production of larger yields through improved varities of crops and intensive cultivation will increase the depletion of nutrients still further. But erosion and leaching cause additional losses. The present production of synthetic nitrogenous fertilizers in the country reached 1.5 million tonnes of nitrogen in 1975-76, and the bulky organic manures might supply another 1.5 million tonnes of total nitrogen. The amounts of phosphorous, potash, etc., added to the soil are very small. It is thus obvious that the current huge drain on nutrient supplies will continue to impoverish the soils unless these supplies are replenished by natural or by artificial means, the principal methods of supplementing natural recuperation and for improving the productive capacity of the soils are:(i) to add organic matter to the soil, so that through decay, it may furnish a more or less continuous supply of nuttrients for crops, and (ii) to restore or increase the amount of defficient nutrients by the application of fertilizers. The urgent need of the constantly expanding agriculture production to meet the requirements of the continually increasing human and cattle populations in India makes the supply of additional plant nutrients through fertilizers and organic manures, a problem of supreme importance.

TYPES OF MANURES AND FERTILIZERS

Indian soils are usually very poor in organin matter as well as in nitrogen. Pohsphate deficiency is less wide-spread and potash deficiency generally occurs in com-pact areas. In acid soils the addition of lome steps up production.Materials which are commonly used to maintain and improve soil fertility

1. *Manures*: These are relatively bulky materials, such as animal or green manures, which are added mainly to improve the physical condition of the soil, to replenish and keep up its humus status, to maintain the optimum conditions for the activities of soil micro-organisms and make good a small part of the plant nutrients removed by crops or otherwise lost through leaching and soil erosion. They, thus, supply practically all the elements of fertility which crops require, though not in adequate proportions. The plant-food elements contained in a manure are released in an available form after it is applied to the soil and is decomposed by soil micro-organisms. Similarly, the green manures add not only substantial amounts of organic matter but also nitrogen.
2. *Fertilizers*: Fertilizers are inorganic materials of a concentrated nature; they are applied mainly to increase the supply of one or more of the essential nutrients, e.g. nitrogen, phosphorous and potash. Fertilizers contain these elements in the form of soluble of readily available chemical compounds. This distinction is, however, not very rigid. In common parlance the fertilizers are sometimes called 'chemical','artificial' or 'inorganic' manures.
3. *Concentrated Organic Manures*: Some of the concentrated materials, such as oil-cakes, bone-meal, urine and blood are of organic origin. The use of manures and fertilizers is complementary and not as a substitute for each other.

4. *Bulky Organic Manures*: The properties and role of organic matter and humus in the soil have been explained already. The average nutrient contents of manures and other organic raw materials which may be used to maintain the humus content of the soil.

Farmyard Manure

Good-quality farmyard manure is perhaps the most valuable organic matter applied to a soil. It is the most commonly used organic manure in India. It consists of a mixture of cattle dung, the bedding, used in the stable and of any ramnants of straw and plant stalks fed to cattle. Though its crop-increasing value has been recognised from time immemorial, more than 50 per cent of the cattle dung produced in the country today is burnt as fuel and is thus lost to agriculture. Not only this tremendous waste, but also the tradition method of preparing and storing the farmyard manure is generally faulty. The cattle-dung, together with stable-waste and house sweeping, is first collected in the open backyard, and when a cartload has been collected, it is removed to another heap or to an uncovered pit in a common plot outside the village. The loose heaps lie exposed to the sun, with the result that the raw organic matter dries up quickly and does not rot properly. Very often, a part of the dry dung is blown off by wind or washed away by rain. Cattle urine is either not conserved or is stored in a defective manner. American studies on the distribution of soil derived elements between urine and faeces of dry cows have shown that 95 per cent pf Potassium, 63 per cent of nitrogen and 50 per cent of sulphur are contained in the urine.

The wastage of nitrogen-rich urine, the loss of nitrogen(in the form of ammonia) due to the fermentation of exposed cattle dung, and the washing away of soluble mineral elements be leaching reduce its manurial value in India to a great extent. Its average content of plant nutrients under Indian and European conditions is shown below for comparison:

Percentage Content

	N	P_2O_5	K_2O
India	0.3	0.15	0.3
European countries	1.0	0.30	1.0

About half of this nitrogen, one-sixth of phosphorous and more than half of potash are readily soluble and subject to dissipation. However, the loss of nitrogen and minerals elements caused by careless handeling can be reduced greatly by using absorbent bedding for cattle, storing dung in stone or brick-line pits, mixing large quantioties of straw and other vegitable matter with cattle dung, and keeping the heap compact and moist. Thus, if urine is properly conserved, the loss of soluble mineral elements through seepage is prevented, bacterial decomposition of raw organic matter is encouraged, plant nutrients are made soluble, and nitrogen losses are minimised.

The relative absorbent capacity of different materials used as bedding for cattle

Material	**Quantity of water (in kg) retained by one kg of the following materials after 24 hours of soaking**
Wheat straw	2.20
Peat straw	2.80
Dry leaves	2.00
Peat	6.00
Sawdust	4.35
Soil	0.50
Sand	0.25

If urine is not conserved in the bedding used for cattle it must be collected in covered *pucca* cistern and then, added to the dung in the manure pit. Nitrogen in the urine is mainly in the form of urea, which readily changes into the highly volatile ammonium carbonate through bacterial action, and quickly loses ammonia thereafter by evapouration. This loss can be reduced to a great deal if the manure and the urine-soaked absorptive itter for bedding are kept compacted in a pit. The pit may be 1 m in depth, 1.3 to 1.5 m in width and 4.5 to 6 m in length, depending upon the no. of cattle on a farm. The filling of the pit should be 'sectional' and when each section of three or 1.3 m in length is filled to about 45 cm above the ground level, it should be clustered with 2.5 cm layer of a mixture of mud and dung in equal proportions. Before plastering, 4 to 5 buckets of water should be added to the manure in the pit. Plastering conserves moisture and nitrogen and also prevents housefly nuisance. The manure becomes ready for use in about 4 to 5 months after plastering.

The quality of manure is also improved by the concentrated feeds given to cattle. Cotton-seed, cotton-seed cake, linseed-meal, wheat bran, grain husk, groundnut cake, gram, horse-gram, etc. are rich in nitrogen, phosphorous, potassium, magnesium and sulphur. It has been found that in the case of adult working-cattle about 80 per cent of nitrogen and the other mineral elements contained in the feed is recovered in urine, faeces and other animal by-products. Accordingly, manure from cattle fed on cereal straws and grass hay is much less valuable than that from animals fed on legume hays, grains and concentrates.

In foreign countries, considerable attention has been given to the use of preservatives on manure. Calcium sulphate or gypsum and superphosphate have proved most promising in preventing the escape of ammomnia. Gypsum has been found specially effective as an ammonia-absorbing agent. Superphosphate, besides absorbing ammonia, supplies additional phosphorous and, thus, improves the crop producing capacity of the manure.

Partially rotten farmyard manure should generally be applied to the soil about three to four weeks before the sowing of a crop. In case there is sufficient moisture in the soil, there sill be enough time for its decomposition and for improving the soil structure. Its application too long before sowing a crop will either cause a drying up of the rotted manure or too quick decomppostion, depending on the incidence of rains. But in each case, there will be a serious loss of ammonia and nitrogen. If the manure is already well rotted, it is advisable to apply it just before sowing to a crop. This procedure is perticularly essential in the case of light soils. In any case, after the manure is carted to the field, it should be evenly spread and worked into the soil soon to avoid the loss of nitrogen. The existing practice of leaving the manure in small heaps scattered in the field for several days, before spreading it in the field and incorporating it into the soil, results in the serious deterioration of its quality, particularly if strong winds blow. In vagitable and fruit cultivation, the application of well-rotten manure, in conjunction with fertilizers to young plants individually, has been founnd to give the best results. In Egypt, even the cotton crop, which is invariably grown on ridges, is given a top-dressing a handful of the rotten manure being applied to the soil at the base of each plant before an irrigation, and is worked into the soil with a hand hoe.

The practice of penning cattle, sheep and goats in the fields in summer is common in some parts of the country. Folding 7,000 sheep for one night is said to add the equivalent of 149.3 quintals(14.93 tonnes) of cattle dung. The fresh dung left in the field in such cases rapidly dries up. This drying checks ammonification and loss of nitrogen. With the first fall of rain, the dung is worked into the soil. It, therefore, does not lose much of its fertilizing value. Further more, its beneficia effects on the physical condition of the soil are undeniable. However, sheep-folding is said to make the land more weedy.

There must be adequate moisture in the soil for the proper decomposition of organic matter. Farmyard manure can, therefore, be applied to all crops grown in the rainy season or grown under irrigation. The quantity of manure to be applied to unirrigated crops varies from 1.5 to 2 cartloads per hectare in areas of heavier rainfall. If sufficient farmy manure is not available, it may be applied at the usual rate to a part of the land,say, to one-third or one-forth of the area, in rotation every year, so that all parts of the field receive the manure regularly once in three or four years. For irrigated field crops, the rate varies from 10 to 20 cartloads. Sugarcane, maize and garden crops, such as potatoes, turmeric, ginger, vegetables and fruits receive still higher doses, amounting sometimes to 15 to 25 cartloads. A cartload of manure, measuring 9 cubic metres, weighs about half a tonne.

It must be stressed that the value of farmyard manure in soil improvement is due to its content of principal nutritive elements and its ability to (i) improve the soil tilth and aeration, (ii) increase the water-holding capacity of the soil, And (iii) stimulate the activity of micro-organisms that make the plant-food elements in the soil readily available to crops. The supply of organic matter, which is later converted into humus is a property of farmyard manure.

One tonne of carttle dung can supply only 2.95 kg of nitrogen, 1,59 kg of phsophoric acid and 2,95 kg of potash.

The use of farmyard manure alone causes an imbalance in nutirtion owing to its relatively low content of phosphoric acid. Therefore, to keep the soils well supplied with all the essential elements of plant food in a readily available form, and also to keep them in good 'heart', it is advisable to use the bulky organic manures in conjuction with superphosphate and such other artificial fertilizers as contain the particular plant food or foods in which a soil may be deficient or which the crop to be grown may specially require.

Composted Manure

Another method of augmenting the supplies of organic is the preparation of compost from farmhouse, and cattle-shed wastes of all types. Composting has been advocated and adopted extensively during the past 25 years. Composting is the process of reducing vegetable and animal refuse(rural or urban) to a quickly utilizable condition for improving the maintaining soil fertility. Research conducted in India and abroad has shown that good organic manure similar in appearance and fertilizing value to cattle manure similar in appearance and fertilizing value to cattle manure can be produced from waste materials of various kinds, such as cereal straws, crop stubble, cotton stalks, groundnut husk, farm weed and grasses, leaves, leaf-mould, house refuse, wood ashes, litter, urine-soaked earth from cattle-sheds and other similar substances. These raw vegetables materials are rich in cellulose and other readily decomposable carbohydrates and have a carbon-nitrogen ratio of 40 or more to 1. The direct application of such undecomposed, low nitrogen organic matter as manure brings about a temporary deficiency of mineral nutrients(specially nitrates and ammonium compounds) in the soil by stimulating the growth of micro-organisms, which in turn, compete with crop plants for available nitrogen, phosphorous and other elements. Hence, before using them as manure, it is necessary to compost or partially decompose them. This process lowers the carbon-nitrogen ratio to about 10 or 12 to 1.

Two methods of composting waste organic materials are usually recommended. One depends on aerobic and other one unaerobic decomposition. In both cases, the farm wastes have to be used as a bedding for cattle in order to absorb a large of the animal urine. In the aerobic process, the used bedding, the sweepings from cattle sheds and some urine-soaked earth from the stable floar are removed everyday, mixed with a little cattle dung and 2 or 3 handfuls of wood ashes are deposited on a well drained site to gradually builed up a lowpile about 30 or 45 cm in height, about 5 m in width and of any convenient length. The pile is built up before to start of rainy season. After the first heavy shower, the waited material in 1.2 strip on each side of the long heap is turn with a rake on to 2.4 m wide stripe in the middle, so raising the height of the heap to nearly 1 m. This process prevents loss of misture and ensures a quick start of decomposition. When the heap sinks appreciably and such a sinking takes about 3 to 4 weeks, it is given a turning and made into fresh heap, thus mixing the outside material with that from inside. After about a month or more, depending on incidence of rains the heap is given a final turning on a cloudy or moderately rainy day and rebuilt the vacant part of the original position. The composed becomes ready for use in about 4 months. This method is eminently suited for composting in the rainy season. The following proposition of raw materials is considered suitable for compost making:

	Parts by weight
Mixed farm residues and cattle-shed	400
Urine-soaked earth	56·
Fresh cattle dung	60
Wood ashes	6

In ·Tamil Nadu, the application of 90 kg of partially fermented dung in the form of a thin suspension, with 22.50 kg of bone-meal per tonne of dry material to be composted is recommended.

If urine-soaked earth and cattle-dung are not available, the raw organic materials can still be composted, provided more than one-third of the residues is soft and finely broken, such as fallen leaves, leaf-mould, kitchen wastes, grass clippings, green and succulent weeds, plant trimming and clopped wheat, barley and soft cereal straw. The use of ordinary soil and wood ashes or lime is, however, essential.

In the anaerobic process, the mixed farm residues are collected in pits of a convenient size, say, 4.5 m*1.5 m*1 m. Each day's collection is spread in a thin layer, sprinkled with a mixture of fresh cowdung(4.5 kg), ashes(140 to 170 g) and water (18 to 22 litres) and compacted. The pit is filled till the raw material stands 38 to 46 cm above its edge, and then is then plastered with a 2-5 cm layer of a mixture of mud cowdung. Under such conditions, decompostion is anaerobic and high temperatures do not develop. Insoluble nitrogen compunds gradually become soluble and the carbonaceous matter is broken down into carbon dioxide and water. The loss of ammonia is negligible, because in high concentrations of carbon dioxide, ammonium carbonate is stable. The plastered pit also prevents the fly nuisance. The well-made compost contains 0.8 to 1 per cent of nitrogen and has all the good properties of farmyard manure. It can be used in the same way as the latter. The anaerobic process is particularly suited for use by gardeners in or near cities and towns.

COMPOSTING METHODS

Raw Materials

The materials are needed are mixed plant residues, animal dung and urine,earth, wood ash and water. All vegetables wastes available on a farm like weeds,stalks,stems,fallen purnings,chaff,fodder remnants,green matter and so on, are collected and stacked in a pile. Hard woody material like cotton or pigeon-pad stalks and stubble are first spread on the farm road and crushed under Vehicles viz. tractors or bullock carts before being piled. Such hard materials should in any case not exceed 10 percent of the total plants residues. Green material which are soft and sufficient and allowed to wilt for 2 or 3 days to remove access moistures before stacking; they tend to pack closely, if they are stacked in the fresh state. While stacking, each material is spread in layers of about 15cm thickness until the heap is about 1m and 50cm high. The mixture of different kinds of vegetable residuesensures a more efficient decomposition.

The heap is then cut into vertical slices and about 20-25 kg is put under the feet of cattle in the shed as bedding for the night. The next morning the bedding, along with the dung and the urine, and urine earth, is taken to the pits where the composting is to be done.

Pit Method

This method includes following steps:

Site and size of pit: The site selected for the compost pit should be near to cattle shed and water source at high level so that no rain water gets in during the mansoon season. A temporary shed amy be constructed over it to protect the compost from heavy raifall. The pit should be about 1 m deep * 1.5-2.0 m wide and of any suitable length.

Filling the pit: The material brought from the cattle shed is spread and on each layer is spread a slurry of dung made with 4.5 kg urine earth and 4.5 kg of inoculum taken from a 15-day-old composting pit. A sufficient quantity of water (nearly 90percent) is sprinkled over the material in the pit to the wet it. The pit is filled in this way layer-by-layer and it should not take longer than 1 week to fill. Care should be taken to avoid compacting the material in any way.

Turning: The material is turned 3 times during the whole period of composting (i) after 15 days from filling the pit, (ii) another 15 days, (iii) after another 30 days. At each turning the material is mixed throughly, moistened with water and replaced with the pit.

Heap Method

During rainy seasons or in regions with heavy rainfall the compost may be prepared in heaps above ground. when sufficient nitrogeneous material is not available a green manure or leguminous crop like sunhemp is grown on the fermenting heap by sowing seeds after the first turning. The green mater is then turned in at the second mixing.

Dimensions

The basic Indore pile is about 2 m wide at the base, 1.5 m high 2 m long. The sides are tapered so that the top is about 0.5 m narrower in width than the base. A small bund is sometimes built around the pile to protect it from wind which tends to dry the heap.

Forming the Heap

The heap isusually commenced with a 20 cm layer of carbonceous material such as leaves, hay-straw, sawdust, wood chips and chopped corn stalks. This is then covered with the 10 cm of nitrogeneous material such as fresh grass, weeds or garden plant residues, garbage, fresh or dry manure or digested sewage sludge. The pattern of 20 cm carbonaceous material and 10 cm nitrogeneous material is followed until the pile is 1.5 m high and they are normally wetted so that they feel damp but not soggy. The pile is sometimes covered with soil or hay to retain heat and is turned at 6- and 12-weeks-interval. In the republic of korea, heaps are covered with thin plasic sheets to retain heat and it has also resulted in the death of insects.

If materials are limited, the alternate layres can be added as they become available. Also, all materials may be mixed together in the pile, if one is careful to maintain the proper proportions. Shredding the material speeds up deconposition considerably; most materials can be shredded by running over it several times with a rotary-mower.

Advantages and Limitations

Preparation on a large scale can be done through community composting. there is lack of protection from rain and wind. A considerable amount of water is needed ans so the heap method is not suitale in near areas of scanty rainfall.

The intense aerobic decomposition to which the material is subjected no doubt shortens the period of composting but it leads to heavy losses of organic matter and nitrogen. Therefore the C:N ratio should be maintained between 30 and 40 to reduce such losses.

Bangalore Method

Preparation of the Pit

Trenches or pits 1m deep are dug; the breadth and length of the trenches can be made depending on the availability of land and the type of material to be composted. The selection of site for 1 pit is made as mentioned in the Indore method. The trenches should preferably have sloping walls and a floor of 90 cm slope to prevent waterlogging.

Filling the Pit

Organic residues and night soil are put in alternate layers and after filling, the pit is covered with a 15-20 cm thick layer of refuse. the materials are allowed to remain in the pit without turning and watering for 90 days. during this period the material settles down due to reduction in volume of the biomass and additional nightsoil and refuse in alternate layers are placed on top and plastered or covered with mud or earth to prevent loss of moisture and breeding of files. the material undergoes anaerobic decomposition at a vey slow rate and it takes about 180-240 days to abtain the finished product.

Advantages and Limitations

The recovery of the finished product is greater as compared to aerobic composting but loss of nitrogen is negligible. Labour requirements is less than for the Indore method as turning of material is not done; labour is needed only for digging and filling the pits.

The methods requires along time to produce a finished compost and so takes up more land use. A uniform high temperature is not assured in the biomass. Problems of odour and fly breeding need to be attended too.

SYNTHETIC COMPOST

In the preparation of synthetic compost, the organic nitrogen as dung, required by micro-organism, can be completely substituted by inorganic nitrogen compounds like ammonium sulphate and urea which are utilized equally effectively for decomposition of carbonaceous materials into compost. The Adco process of preparations of synthetic compost developed by Hutuchinson and richards is based on this principle.

This facilitates the utilization of large quantities of various organic waste material where supplies of dung are either short of the requirement or not available at all, as on mechanized farms.

The basic principle of C:N ratio in manure preparation can be applied to add nitrogenous fertilizers in sufficient quantity to decompose. The material to be composted is moistened. This is sprinkled with the fertilizer solution and then with lime. Superphosphate may be added to fortify the phosphorous content of the manure. The treatment is continued layer-wise until the heap or pit is filled to the size and allowed to ferment. The manure becomes ready for application in about 120-180 days and resembles with farmyard manure in its action on soil and plant growth.

Leaf Compost

Leaf composting, can be achieved by heap or ditch composting or by windrow composting. windrow are preferred as they allow efficient handling of materials Provide good aeration, allow sufficient of water and are easy to be performed.

It is suggested the formation of uniform-shaped windrow from 2.40 m-3.60 m at the base and 2.40m-3.00m high and of any convenient lenght. Windrow built too high will have excessive compaction at the base resulting in anaerobic conditions. Windrows built too low will not allow sufficient insulation to sustain thermophilic temperatures during cold weather. To ensure proper aeration, it is important to break apart tightly compacted leaves.

Though reduction in size may aid in rapid decomposition, it is not desirable in leaves because it increases the compaction making more frequent aeration necessary. When the incoming leaves are not adequately moist, it is desirable to add water to maintain a proper moisture regime of 40 to 60percent.

The C:N ratio of leaves is relatively high. It can be as high as 80, and needs to be amended with nitrogen. Sewage sludge, urea and grass clippings are good sources of nitrogen. If a nitrogen source is to be added, caution should be exercised to distribute it uniformly throughtout the windrow, lest it may result in undesirable anaerobic conditions and uneven decomposition. Proper aeration can be maintained by periodical mixing of the material.

Under optimum environmental composting conditions, leaf compost will be ready between 180 and 270 days. Leaf mould will have final P^H range of 6-7.

It is suggested the use of finished compost (leaf mould) as converting material (10-15cm) in the subsequent preparation of leaf compost to supply a heavy inoculum of micro-organisms.

Enrichment with Phosphorous

Phosphorous enriched compost is prepared by adding 5percent superphosphate at the filling of the compost pits. Other sources of phosphorous for this purpose are powered rock phosphate, preferably of low grade (less than 11percent P), can be used with profit. Besides, phosphorous it is a source of calcium and micronutrients. Bonemeal will provide nitrogen as well as phosphorous; it contain 9-11percent P and 2-4percent N. Steamed bonemeal is more easily ground than the fresh material. It contains a little less nitrogen but more phosphorous than the raw material. Basic slag provides calcium, magnesium and trace nutrients and small amount of phosphorous. Banana residues contains about 1-1.5percent phosphorous on an ash basis.

Enrichment with Potassium

Granite dust of powdered potassium-containing minerals like feldspars can be added to enrich compost. Potassium and other deficient elements can be added to compost by including plant materials which contain apperciable amounts of those elements. Water hyacinth is a rich source of potassium and of many other elements required by plants. Banana skin and stalks contain 34-42percent potassiumon an ash basis, seaweeds are rich in iodine, boron, copper, magnesium, calcium and phosphorous. Leaves are also a good source of trace elements and should form a parta part of every compost heap. Potato peel is rich in trace elements and dry potato vines contain 1percent potassium, 4percent calcium and 1percent magnesium.

Vermi Composting

Vermi-composting is the use of earthworms for composting of organic residues. Earthworms can consume practically all kinds of organic matter. One worm, weighs about 0.5 to 0.6 g, eates wastes as their own body weight perday and produces cast of the same weight per day. It is estimated that 1000 tonnes of moist organic matter can be converted by earthworms into 300 tonnes of compost.

Organnic materials undergo comples biochemical changes in the intestines and vermi-composting is an appropriate technique for disposal of non-toxicsolid and liquid organic wastes. It helps in cost effective and efficient recyclicing of animal wastes (poultry, equine,piggery excreta and cattle dung) agriculture residue and industrial wastes using low energy, excreta together with their cocoons and undigested feed make up vermicasting.

The castings of earthworms are rich in nutrients (N, P, K, Ca and Mg), and also in bacterial and actinomycetes population. The actinomycetes population in worm cast is over 6 times more than in the original soil. A mosist compost heap (30-40percent moisture level) of 2.4 m * 1.2 m * 0.6 m high, can support a population of mare than 50,000 worms. The temperature is the culture bed should be within the range of 20^0-30^0C. The introduction of worms into compost heap has been found to mix the materials, aerate the heaps and hasten decomposition. turning the heaps is not necessary, if earthworms are present to do the mixing and aeration. Besides rural and urban wastes, effluents from agro-industries viz dairies, tanneries, pulp and paper mills, distilleres etc.can be treated by using earthworms.

Benefits of Earthworms

Earthworms help in the preparation of compost maintaining soil health as follows:

1. Improvement in fertility of soil.
2. Arnelioration of physical condition of soil.
3. Mixing of sub-soil and top soil.
4. Correction of undetermined deficiences in plants.
5. Use of earthworms in recycling of city and rural wastes, sewage waste waters and suldge, and industrial wastes eg. paper, food and wood industries.
6. Supplementing traditional feeds

Species for Vermi-composting

Earthworm can be divided as surface living (epigeic) and burrowing (epianecic) worms. Epigeic or compost worms are found on surface and are reddish brown eg. Lumbricus rubellus (red worms). Of many species of earthworms tested for mass culture all over the world, Eisenia fetida, Eudrilus eugeniae and Perionyx excavatus come in the above order of preference for their ability to compost organic wastes. The shapes of cocoons of Eisenia frtida and Eudrilus eugeniae are dissimilar.

Rearing of Earthworms

The worms are reared and multiplied from a commercially obatined breeder stock in shallow wooden boxes of 45 cm *60 cm, provided with drainage holes and stored on shelves and tiers.

A bedding material is compounded from miscellaneous organic residues saw dust, cereal straw, rice husks, sugarcane trash, bagasse, paper, cardboard, coir waste, grasses etc. and is moistened well with water. The wet mixture is stored for 30 days covered with a damp sack and is throughly mixed several times. When fermentation is complete, chicken manure and green matter eg. Leucaena leaves or water hyacinth is added. The material is placed in the boxes in the ans sufficiently loose for the worms to burrow and should be able to retain moisture. the The proportion of the different materials will vary according to the nature of the material but a final nitrogen content of about 2.4percent should be amied at. A pH value as near neutral as possible is necessary and the boxes should be kept at temperatures between 20^0C and 27^0C. At higher temperatures the worms will aestivate and at lower temperatures they hibernate. for each 0.1 m^2 of surface area 100 g of breeder worms are added to the boxes. Inspite of their being able to eat the bedding material, the worm at this stage are regularly fed @ 1 kg of feed a day for every kg of worms. the feed stuffs used are again various types of organic matter and include partially digested cowdung, chicken manure,

Leucaena leaves, vegetable waste and water hyacinth. Some form of protection against predators like birds, vats, ants, frogs, leeches and centipeded is provided to the worms.

Vermi-composting in Pits

A number of pits 2m *1m with sloping sides aredug having suitable dimension. Vermi-composting is done in pits and in vitro. Both of these are discussed here.

Bamboo poles are laid in paralled row on the pit's. Its floor is with a lattice of wood strips. Necessary drainage is provided because worms can not survive in a waterlogged condition. Alternatively to this and sand can be placed in the bottom of the pit to facilitate proper drainage. Above this a thick layer (15-20 cm) of good loamy soil should be spread. The pit can now be filled with available organic residues such as animal manure, leaves and green weeds, crop residues etc. Moisture levels of the contents of pit is maintained through addition of required amount of water. The worms from breeding boxes are introduced in the organic refuse, the worms immediately burrow down into the damp soil.

The compost pit is left for 60 days. It should be shaded from hot sunshine and it must be kept moist. Within 60 days about 10 kg of castinge would have been producer per kg of worms. The pit is then excavated to an extent of about two-thirds to three-quarters and the bulk of the worms removed by hand or by seving. This leaves sufficient worms in the pit for further composting and the pit can be refilled with fresh organic residues and continued.

The compsost can be sun-dried and sieve to give a good quality compost. The average nutrient content of vermi-compost is N 0.6-1.20percent, P_2O_5 1.34-2.20percent, K_2O 0.4-0.67percent, CaO 0.44percent and MgO 0.15percent.

The excess worms that have been harvested from the pit can be used in the other pits, sold to other farmers for compost inoculation, and may be used as animal and poultry feed or fish food.

Method of Pit Vermi-composting

Selection of earthworm: Earthworm which is native to the local soil and vermi-compost may be used.

Size of pit: Any convenient dimensions such as 2m * 1m *1m may be prepared. This can hold 20,000-40,000 worms giving 1 tonne manure/month.

Preparation of vermibed: a 15-20 cm thick layer of good loamy soil above athin layer (5cm) of broken bricks and sand should be made. This layer is inhabited by earthworms.

Inocluation of earthworms: About 100 earthworms are introduced as an optimum inoculating density into a composit pit of about 2m * 1m * 1m provided with a vernibed.

Organic layering: It is done on the vermibed with fresh cattle dung. The compost pit is then layered to about 5 cm with dry leaves or hay. Moisture content of the pit without flooding is maintained through the addition of water.

Wet organic layering: It is done after 28 days with moist/ green organic waste which can be spread over it to a thickness of 5 cm. This practice can be repeated every3-4 days. Mixing of wastes periodically without disturbing the vermibed ensures proper vermi-composting. Wet layering with organic waste can be repeated till the compost pit is nearly full.

Harvesting of vermi-compost: At maturation, the moisture content is brought down by stopping the addition of water for 3-4 days. This ensures drying of compost and migration of worms into the vermibed. The mature compost, a fine loose granular mass is removed out from the pit, dried and packed.

Rate of application: Mature vermi-compost is recommended @ 0.5 tonnes/ha.

To boost vermi-compost production production following suggestions should be followed:

a. A mixture of cattle, sheep, horse dung with gram and wheat bran and vegetable wastes forms the ideal feed for worms.

b. Mixing of gram bran with dung mixture in 3:10 ratio increase the biomass.

c. Mixing of wheat bran to dung mixture in 3:10 ratio hastens the growth of worms. Addition of kitchen waste in the same proportion increases the worm population.

d. The biogas sludge and poultry dropping in equal quantities enhance the worm population and the biomass.

In-vitro Vermi-composting

This is also called as bioconversion in soil. This involves the application of the basal dose (5 tonnes/ha) of vermicastings and covering with 2.5 cm. Layer of organic mater (cowdung or pressmud) followed by 10 cm layer of sugarcane trash, crop residues or city swastes. The worms hatch out within 10 days.

Vermi-composting of Agricultural Wastes

With the aim of vermi-composting of agricultural wastes an experiment was conducted at the Indian Agricultural Research Institute, New Delhi. The usefulness and efficiency of earthworms in composting of agricultural residues was studied.

Two trials were conducted with mixed organic materials. Trial A, had more of green materials like grasses and Leucanena leaves mixed with soil and paper. Experient B had 4 kg of composting materials consisting of 2 kg of paddy straw, 1 kg soil 500 g twigs, and 500 g shredded paper and was laid in layers per pit. The bottom most layer was laid with twigs to allow for percolation of excess moisture. The compost pit was kept moist and care was taken to avoid waterlogging.

Town Compost

In recent years, large-scale composting of town refuse and night-soil in properly constructed trenches away from human habitations has been taken up successfully by the municipalities of many large and small towns. Trenches, 1 to

1.2 m wide, 75 cm deep and of convenient length, are filled with successive layers of night-soil, town refuse and earth, in this order. The compost gets ready in about three months. The following figures of volume-weight conversion will be found useful in the preperation of town compost.

Volume		Weight in kg or q
1 cu.m of refuse	=	318 kg or 3.18 q
1 litre of night-soil	=	0.991kg
1 cu. m of compost	=	636 kg or 6.36 q
1 cartload of refuse	=	95.40 kg

Social prejudice against the use of this valuable compost has disappeared and town-composting is almost being rapidly adopted in other localities all over India. With a suitable modifications, such as the provision of trench latrines, it can be taken up in villages too.

Sewage and Sludge

The liquid waste, like sulage and sewage contain large quantities of plant nutrients and are used for growing of sugarcane, vegetables and fodder crops near many large towns by operating sewage-farms. In many places, the undiluted sullage has been found to be too strong for healthy plant growth and if it contains readily oxidized organic matter, its use actually reduces nitrates present in the soil. The disadvantages are still greater if sewage is used on land without preliminary treatment. The soil quickly becomes 'sewage sick' owing to the mechanical clogging by colloidal matter in the sewage and the development of anaerobic organisms which not only reduce the nitrates already present in the soil but also produce alkalinity. Bacterial contamination makes the eating of raw vegetables on untreated sewage a real danger to health.

For these reasons, it is now usual to construct a setting or a septic tank in which sewage is stand allowed to relieve it of the heavier portion of the solid matter in it, or to undergo a preliminary fermentation. The effluent from the settling-tank, however, still carries a large amount of objectionable colloidal matter, and the deposit of sludge that settles in the tank is of small manurial value and is often offensive. These defects can be removed by thoroughly aerating the sewage in the settling-tank by blowing air through it. The sludge that settles at the bottom in this process is called 'activated sludge' It has the remarkable property of bringing about the rapid oxidation of the organic matter present in fresh sewage. It is also in offensive and on dry-weight basis, contains 3 to 6 per cent N, about 2 percent P_2O_5 and 1 per cent K_2O in forms that can become readily available when applied to the soil. Similarly, the effluent is a clear, odourless liquid containing nitrates in solution, from shich most of the pathogenic bacteria originally present have been removed. Both the activated slidge and the effluent can be used with safety for manuring and irrigating crops. However, under no circumstances should any produce grown on a sewage-farm be eaten uncooked.

Night-soil or Poudrette

Very few towns India are equipped with complete sewage. The sanitory disposal of night-soil with an effective control of foul smell and fly nuisance is, therefore, a serious problem all over the country. Since human excrements are a potential source4 of soil improvement, public health authorities in many towns make the necessary arrangements for its conservation and conversion into a form in which it can be safely used as a manure. The dehydration of night_soil, as such, or after admixture with absorbing materials,e.g. soil, ash. charcoal and sawdust, produces a poudrette that can be easily used as a manure. The mixing of night-soil with an equal volume of ash and 10 per cent powdered charcoal produces an odourless material,containg 1.32 per cent potash and 24.2 per cent time. The addition of 40 to 50 per cent of sawdust to the night-soil yields straightway a dry, acidic poudrette which may contain 2 or 3 per cent nitrogen. The annual potential quantity of manurial ingredients in the night-soil from India's present population of 600 millions is estimated at appromimately 8.1

million tonnes of dry matter containing almost 0.4 million tonnes of nitrogen, 0.25 million tonnes of phosphoric acid and 0.17 million tonnes of potash.

Green Manures

Despite special efforts at increasing the supplies of farmyard manure and compost, the supply of farm and other organic manures is scarce and ever becoming more costly. Green-manuring, wherever feasible, is the principal supplementary means of adding organic matter to the soil. It consists in the growing of a quick-growing crop and ploughing it under to incorporate it into the soil. The green-manure crop supplies organic matter as well as additional nitrogen particularly if it is a legume crop, which has the ability to acquire nitrogen from the air with the help of its root-nodule bacteria. A leguminous crop pròducing 8 to 25 tonnes of green matter per hectare will add about 60 to 90 kg of nitrogen when ploughed under. Rhis amount would equal an application of three to ten tonnes of farmyard manure on the basis of organic matter and its nitrogen contribution. The green manure crops also exercise a protective action against erosion and leaching.

The crops most commonly used for green-manuring in this country are the following:

Sunnhemp *(Orotalaria juncea)*, dhaincha*(Sesbania aculeata)*, cluster-bean*(Cyamopsis tetragonoloba)*, senji*(Melilotus parviflora)*, cowpea*(Vigna catjang, V. sinensis)*, horse-gram*(Dolichos biflorus)*, pillipesara*(Phaseolus trilobus)*, berseem or Egyptian clover*(Trifolium alexandrinum)*.Lentil*(Lens esculenta)* is recommended in Kashmir for green-manuring paddy. Sown in late autumn, it is said to provide a winter cover and make new growth in early spring for ploughing under before the the sowing of peas in the standing crop of irrigated cotton in Uttar Pradesh, and the sowing of sunnhemp in the standing paddy crop in Andhra Pradesh, Tamil Nadu and Karnataka states, or of *val IDolichos lablab)* in the coastal paddy areas of Maharashtra, are valuable practices for soil improvement. All these leguminous crops leave the soil in a better physical conditions and richer in nitrogen. The growing of pulses mixed with cereals all over India, the intercropping of cotton with groundnut of with *tur(Cajanus indicus)* in central and southern parts of the country, or with cluster-bean, *mung and moth(Phaseolus aconitifolus)* in Punjab and adjoing states; the growing of wheat mixed with peas and gram in northern and central India; and the growing of fodder sorghum mixed with *Dolichos lablab* in some parts of Tamil Nadu also enrich the soil.

In localities near forests in Tamil Nadu, Karnataka and Andhra Pradesh, the paddy crop is often manured with green forests leaves. These are incorprated into the soil at the time of puddling. In recent years, extensive efforts have been made in these states to plant *Glyricidia maculata* and *Sesbania speciosa* on the borders of paddy fields or in other vacant spaces to provide green leaf for manuring the paddy crop. Grown from seedlings or rooted stumps, planted 2 m apart, each *Glyricidia* plant is said to give annually tow cuttings each of about 6 to 12 kg of green leaf. It does well in both red black soils. Similarly *Sesbania speciosa* seedlings planted 10 cm apart on paddy borders produce 1,000 to 2,500 kg of green leaf for manuring 0.4 ha of paddy. Only 115 g of seed is needed to provide seedlings sufficient for border-planting ot two hectares of paddy. In certain other paddy-growing areas, *Pongamia pinnata(karanji), Tephrosia, Terminalia* and other trees yielding large quantities of leaves are planted for use as a green manure. In the Malabar districts of Tamil Nadu, *Indigofera teysmanni* is grown to pfovide green leafy twigs for manuring paddy.

For the proper rotting of the green manure, it is necessary that the green material should be succulent and there should be adequate moisture in the soil. Plant at the flowering stage, contain the greatest bulk of succulent organic matter with a low carbon/nitrogen ratio. The incorporation of the green-manure crop into the soil at this stage allows a quick liberation of nitrogen in the available form. With advancing age, the percentage of carbonaceous matter in the plants increases and that of nitrogen decreases. if the material with a side carbon/nitrogen ratio isploughed under, micro-organisms bring about its decomposition, draw upon the released nitrogen and mineral nutrients and cause a temporary nutirient deficiency.

Sometimes *methi* or *dhaincha* is sown in-between the rows of the newly planted sugarcane crop, or cluster-bean is planted between the rows of irrigated American cotton. When the leguminous plants are five to six weeks old, they are incorporated into the soil. Sugarcane and cotton are claimed to benefit from the legumes. The decay of green matter in this case seems to occur when it best serves as a fertilizer for the beneficiary crop.

The increase of yield after green-manuring is usually of the order of 30 to 50 per cent. The fertilizing value of the legume crop can be increased a great deal by manuring it with superphosphate. This practice not only increases the phosphorous content of the green-manure plants, but also encourages their plant growth on the whole, thus converting an inorganic fertilizer into an organic manure. Green manures have a marked residual effect also.

FERTILIZERS

Despite the special steps taken in recent years to increase the supply of farmyard and other bulky organic manures, the available quantities are insufficient to meet the existing and prospective needs. Fertilizers have also the dvantage of smaller bulk, the resultant easy transport, relatively quick availability of their plant-food constituents and the possibility of their application in proportions suited to the actual requirements of different crops and soils. Fertilizers are usually classified according to the three principal elements and may, therefore, be included in more than one group.

Nitrogeneous Fertilizers

According to the manner in which their nitrogen is combined with other elements, the nitrogenous fertilizers are divided into four groups; nitrare, ammonia and ammonium salts, chemical compunds containing nitrogen in the amide form, and plant and animal by-products. Sodium Nitrates:

It is also known as 'Chilean' nitrate. It owes its importance as a pioneer nitrogenous fertilizer. It occurs in natural deposits in northern Chile and is refined before shipping. The refined product contains about 16 per cent nitrogen in the nitrate form, which renders it directly available tp plants.

For this reason, it is highly valued as a source of nitrogen when applied as top-and side-dressings, especially to young plants and garden vegetables, which need readily available nitrogen for quick growth leached out from the soil. for wheat, maize, barley, cotton, sugarcane, etc., it is as benefecial as ammonium sulphate.

Sodium nitrate is particularly useful for acidic soils. Its continued and abundant use in soils is said to cause deflocculation and develop a bad physical condition in regions of low rainfall. It should be stored in a dry ware-house

Ammonium Sulphate:

It is the most widely used fertilizer in the country. It is a white crystalline salt, containing 20 to 21 percent aoomniacal nitrogen. It is easy to handle and it stores well under dry conditions. During the rainy season, it sometimes, forms lumps. These lumps should be powedered before use. Being soluble in water, it acts quickly, but despite its high solubility, its nitrogen is not readily lost in drainage, because the ammonium ion is retained by the soil particles. it is, therefore, very suitable for wet-land crops, e.g. paddy and jute. It has also been found useful on wheat, cotton, sugarcane, potatoes and many other crops grown on a wide variety of soils. It has, however, an acid effects on the soils. Its long-continued use increases soil acidity and lowers the yield. The application of this fertilizer to acid soils improved the yield of tea plants considerably. It is advisable to use this fertilizer in conjunction with bulky organic manures to safeguard against the ill effects of continued application of ammonium sulphate to field and horticultural crops.

Ammonium sulphate can be applied before sowing, at sowing time, or as a top-dressing to the growing crop. it should not be applied along with, or too close to, the seed, because in concentrated form, it affects seed germination very adversely.

Ammonium Nitrate:

It is a white crystalline salts, containing 33 to 35 per cent nitrogen, half as nitrate nitrogen and half in the ammonium form. in the ammonium form, it cannot be easily leached from the soil. This fertilizer is quick-acting, but highly hygroscopic and not fit for storage. Granulation of the material and a light coating of the granules with oil reduce hygroscopicity to some extent. It has an acidulating effect on the soil. Under certian conditions, it is explosive, and, therefore, it should be handled cautiously.

'Nitro Chalk' is the trade name of a product formed by mixing ammonium nitrate with about 40 per cent lime-stone or dolomite. It is granulated, non-hazardous and less hygroscopic. it contains 20.5 per cent nitrogen, half in the form form of ammonia and half as nitrate. The presence of lime in it makes it particularly useful for acid soils.

Ammonium Sulphate Nitrate: 3

It is a mixture of ammonium nitrate and ammonium sulphate. it is available in a white crystalline form or as dirty-white granules. This fertilizer contains 26 per cent nitrogen, three-fourths of it in the ammoniacal form and the rest(6.5 per cent) as nitrate nitrogen. It is non-explosive and not as deliquescent as ammonium nitrate. It is readily soluble in water and is very quick-acting. its keeping quality is good and it is useful for all crops. Its keeping quality is good and it is useful for all crops. Its acid effect on the soils is only one-half of that of ammonium sulphate. It can be applied before sowing, at sowing time or as a top-dressing, but it should not be applied along the seed.

Ammonium Chloride

It is a white crystalline compound, possessing a good physical condition and containing 26 per cent ammoniacal nitrogen. It is extensively used on paddy in Japan, In India, it is used largely in industries. In general, it is similar to ammonium sulphate in action. It is usually not recommended for tomatoes, tobacco and such other crops as may be injured by chlorine.

Urea

It is a white, crystalline, organic chemical. It is a highly concentrated nitrogenous fertilizer, containing 45 to 46 per cent of non-proteined organic nitrogen. it is fairly hygroscopic and presents considerable difficulty, it is also produced in granular or pellet forms and is coated with a non-hygroscopic inert material. It is highly soluble in water and, therefore, subject to rapid leaching. It is, however, quick-acting. When applied to the soil, its nitrogen is rapidly changed into ammonia. Like ammonium nitrate, urea supplies nothing but nitrogen.

It may be applied at sowing time or as a top-dressing, but should not be allowed to come into contact with the seed. It is suitable for most frops and can be applied to all soils.

Ammonia

It is a gas containing about 80 percent of nitrogen. Under suitable conditions of temperatures and pressure, it becomes liquid(anhydrous ammonia). Another form, 'aqueous ammonia', results from the absorption of ammonia gas into water, in which it is soluble. Ammonia is used as a fertilizer in both these forms. Anhydrous ammonia can be applied by introducing it into irrigation water, or directly into the soil from special containers, which makes its use rather expensive. Its possibilities as manure for paddy, sugarcane and cotton are being onvestigated at Bangalore in the Karnataka state. In manurial experiments on cotton in Maharashtra, aqueous ammonia has been found to be as efficient as ammonium sulphate.

Calcium Ammonium Nitrate

Calcium ammonium nitrate is a fine free-flowing, light brown or grey granular fertilzer. It is commercially prepared from ammonium nitrate and ground limestone. It is almost neutral and can be safely applied even to acid soils. Its total nitrogen content may vary from 25 to 28 per cent. Half of this total nitrogenis in the ammoniacal form and half is in nitrate form. According to the prescribed standards, its moisture content should not be more than 0.5percent by weight. As regards the particle size, 90 per cent of the material should pass through 4 mm IS sieve and be retained on a 1-mm IS sieve but not more than g per cent shall be below 1 mm.

Organic Nitrogenous Fertilizers

These fertilizers include plant and animal by-products, such as oil cakes, fish manure and dried blood from slaughter houses. Before their organic notrogen can be used by the crops, it is converted through bacterial action into readily usable ammonia-nitrogen and nitrate nitrogen. These fertilizers are, therefore, relatively slow-acting, but they supply available nitrogen for a longer perios. Furthermore, they may also small amounts of organic stimulants that they may contain, or of some of the minor elements needed by plant.

Oil-cakes of different kinds are produced in India to the tune of about two million tonnes anually. They contain not only nitrogen but also some phosphoric and potash, besides a large quantity of organic matter. The chemical composition of the principal oil-cakes available in the country. In addition to the three fertilizing constituents (N, P_2O_5 and K_2O), the oil-cakes invariably contain2 to 15 per cent of oil, depending on whether the oil is extracted by using solvent process or with expelers, hydraulic presses or indigeneous *ghanis*. This residual oil, however, dous not, in practice, affect their manurial value. Owing to the great importance of using edible oil-cakes as a cattle feed, their utilization as fertilizers is undesirable. They should be fed to cattle and the excrements may be used as manure. Inedible cakes, like castor cake, *neem, mahua* cake and *karanj* cake can, however, be recommended for use in conjunction with quicker-acting chemical fertilizers.

Dried blood or blood_meal contains 10 to 12 per cent highly available nitrogen and 1 to 2 per cent phosphoric acid. It is a very quick-acting manure and effective on all crops and all types of soils. It should be used in the same way as oil cakes.

Fish manure is available either as dried fish or as fish-meal or powder. In regions where fish oil is extracted, the residue can be used as a manure. Depending on the type of fish, its manurial constituents vary from 5 to 8 per cent of organic nitrogen and from 4 to 6 per cent of phosphoric acid. It is quick-acting and suitable for all crops and soils. It should preferably be powdered before use.

Phosphate Fertilizers

These are classified as natural phosphates, treated or processed phosphates, and by-product phosphates anc chemical phosphates.

Rock Phosphate

It occurs as natural deposits of rock in Morocco, the United States of America, Poland, Russia, Tunisia, Algeria, Algeria, Brazil, Egypt, Nehru and some islands in the Pacific Ocean and the Indian Ocean. It contains 25 to 35 per cent phosphoric acid, but this phosphorus is insoluble in water. Practically no rock phosphate, as such, is used as a fertilizer in this country, except some quantity in southern India. In other countries, finely pulverized phosphate rock has been found to give satisfactory results in soils which are very deficient in phosphprus and are acidic. Adequate rainfall and a long growing period of the crop enhance the response. All the same, very little rock phosphate is used is directly as a fertilizer even in these countries. Much more of it is used to manufacture superphosphate, the phosphoric acid of which is water-soluble and is in an available form.

Superphosphate

It is the most widely used fertilizer in India. Prepared formerly by treating bones with sulphuric acid, it is now manufactured largely by treating ground phosphate rock with almost an equalquantity by weight of sulphuric acid. This treatment produces a brownish-grey mixture containing monocalcium phosphate and calcium sulphate(gypsum) in practically equal quantites. This fertilizer is manufactured in three grades: single superphosphate containing 16 to 20 per cent phosphoric acid; dicalcium phosphate, 35 to 38 per cent; and triple superphosphate, 44 to 49 per cent. Single superphosphate is the most commonly available grade in the Indian market. Triple superphosphate in the prodiction of which liquid phosphoric acid is used instead of sulphuric acid, contains very little calcium sulphate. Tripple superphosphate is used mostly in the manufacture of concentrated mixed fertilizers.

The phosphoric acid in superphosphate is wholly water-soluble but when applied to the soil, it is immediately converted into insoluble phosphate owing to precipitation as calcium, iron or aluminium phosphate, according as the soil is alkaline or acid.

Thus athe fertilizer is not leached, but is slowly dissolved in the soil solution. Fixation losses can be reduced by applying the fertilizer in bands on both sides of the row of seeds at a depth of 10 to 15 cm with a drill. By this means, at least some of the phosphate does not come into direct contact with the soil and thus the available phosphorous is readily released for absorption by the plant roots.

The fertilizer is suitable for all crops and can be applied to all soils. In acid soils, it should properly be used in conjunction with organic manure. It should be applied before or at sowing or transplanting.

Basic-slag

This is a by-product of steel factories. Depending on the phosphorous content of the iron ore, it contains from 6 to 20 per cent of phosphoric acid (P_2O_5). Slag from Indian steel-mills is poor in P_2O_5 and is not used as a fertilizer. The high-grade European slag containing 15 to 18 per cent P_2O_5 is a popular phosphatic fertilizer in central Europe. It is not as soluble as superphosphate, but unlike the latter, it is alkaline in reaction and good for acid soils. For effective use, it must be pulverized before application.

Bone-meal

The use of raw bones as amanure for fruit_trees is an age-old practice in this country. Burying the skeleton of an animal under a fruit-tree is known to benefit its growth and bearing. Large quantities of bones used to be exported until a few years ago. Exports have since declined and bone-meal(i.e. ground bone) is now a widely used phosphate fertilizer. It is available in two forms:

a. Raw bone-mel, and

b. Steamed bone-meal.

The steaming of bones under pressure removes fats, greases, nitrogen and glue-making substances. Thus, while raw bone-meal contains about 4 per cent slow-acting organic nitrogen and 20 to 25 per cent insoluble phosphoric acid. Steamed bones are more brittle and can be readily ground. This is an advantage, as the rate of availability of phosphoric acid. Steamed bones are more brittle and can be readily ground. This is an advantage, as the rate of availability of phosphoric acid in bones depends largely on their degree of pulverization. Bone-meal contains only 1 to 2 per cent nitrogen but 25 to 30 per cent phosphoric acid. Steamed bones are more brittle and can be readily ground. This is an advantage, as the rate of availability of phosphoric acid. in bones depends largely on their degree of pulverization. Bone-meal, having particles not larger than3/32 inch, considered suitable for use as a fertilizer, but the more finely powedered it is, the quicker its P_2O_5 becomes available in the soil.

Being relatively slow-acting, bone-meal should not be used as a top-dressing; it must be incorporated into the soil in order to become available. It may be applied either at sowing time or a few days before sowing and should be broadcast. It is particularly suitable for acid soils. It is considered a safe manure for all crops.

In some parts of the country, charred and powedered bones are used as a manure. Charring destroys about half the nitrogen, but leaves intact practically the whole of P_2O_5 in a quickly available form. In the absence of arrangements for steaming and grinding, charring can be easily carried out even in remote villages.

Potassic Fertilizers

Most of the Indian fertilizers soils contain a sufficient amount of potash. Potassic fertilizers should, therefore, be applied only to such soils as are definitely known to be defecient in potash or to those which respond to their application, such as sandy soils. They can also be applied to certain crops, such as tobacco, potato, onion, tomato and fruit-trees, to improve the quality and appearance of their produce.

Potassic fertilizers in common use are:

a. Muriate of potash(potassium chloride), and

b. Sulphate of potash (potassium sulphate).

These salts are important constituents of the waters of oceans and inland seas and of saline deposits derived thereform. The largest known deposits of these salts are at Stassfurt in Germany, in the Caspian Sea region in Russia, in the Dead Sea in Palestine and at some places in California, New Mexico, france and Spain.

Muriate of Potash

It is a gray crystalline material containing 50 to 63 per cent potash (K_2O), the whole of which is readily available. Though highly soluble in water, it is not lost from the soil, as it is absorbed on the colloidal surfaces. it can be applied at sowing time or before sowing.

Sulphate of Potash

It is made by treating potassium chloride with magnesium sulphate and is, therefore, more costly. it contains 48 to 52 per cent K_2O. It dissolves readily in water and becomes available to the crop almost immediately. It can be applied at any time up to sowing, but should not be drilled with the seed. It is considered better than muriate of potash for crops, such as tobacco, chillies, potato and fruit-tree, where quality is of prime importance.

Other Sources

Wood ashes, cattle-dung ash, leaf-mould, tobacco stems and water hyacinth are the available indigenous sources of potash. Unleached wood ash contains 5 to 6 per cent of potash in the form of potassium carbonate (which is alkaline), 1 to 2 per cent phosphoric acid and 25 to 30 per cent lime (Cao). Both potassium carbonate and lime in the wood ashes counteract the acidity in the soil. Groundnut shell, paddy husk and bagasse ashes are available near decorticating factories and rice and sugar-mills. These also contain a fair amount of potash and some phosphoric acid. Ground tobacco stems contain 2 to 3 per cent of nitrogen and 6 to 10 per cent of potash in quickly available forms. Hyacinth abounds as a weed in fresh-water ponds in Bengal, Assam, Tripura and Malabar in Kerala. When dry, it contains 1 per cent nitrogen, 4 per cent potassium and a small quantity of phosphorous.

Soil Amendments

Lime is generally used for correcting soil acidity, for improving the physical condition of the soil and encouraging bacterial activity. Similarly, gypsum is used for reclaiming 'alkali' soils or land from the sea and improving the structure of heavy black clay soils. Hence, these are called soil amendments.

Compund Fertilizers

These fertilizers are multiple nutrient materials, supplying two or three plant nutrients simultaneously. When both nitrogen and phosphorous are deficient in a soil, a compound fertilizer, e.g. ammophos, can be used. It contains 16 per cent N and 20 per cent P_2O_5. Its use does away with the necessity of purchasing two different fertilizers and mixing them in correct proportion before use. The nutrient contents of some of the other compound fertilizers are shown below:

	N	P_2O_5	K_2O
Monoammonium phosphate	11.0	48.0	-
Diammonium phosphate Ammoniated	2.0 -	14.0 -	-
superphosphate(4 - 7 per cent)	4.0 -	20.0	-
Potassium nitrate	13.0	-	44.0

Mixed Fertilizers

Compound fertilizers contain plant food elements in fixed propertions and are, therefore, not always best adapted ti different kinds of soils. Accordingly, the needs of different soils can generally be met most economically by the use of fertilizer mixtures containing two or more materials in suitable propertions. Mixtures usually meet nutrient deficiences in a more balanced manner and require less labour to apply than straight fertilizers used seperately. Mixtures containing all the three principal used seperately. Mixtures containing all the three principal nutrients (N, P and K) are termed complete fertilizers.

Some manufacturers prepare special mixtures for different crops, such as wheat, sugarcane, pady, potatoes, tabacco, fruit-trees and vegetables. Trade names like Ammophos, Niciphos and Nitrochalk are employed for other proprietary products, In foreign countries, even insecticides, fungicides and weed-killers, such as DDT, BHC and mercury or copper salts and 2, 4-D are sometimes incorporated into fertilizers mixtures.

When a required mixture is not available of the cost of the mixture sold in the market is high in relation to the cost of its individual components, it may be made at home by mixing the constituent fertilizers in correct proportion. The preparation of mixed fertilizers requires a good knowledge of the properties and mutual reactionof the component straight fertilizers under different climatic and storage conditions. Therefore in preparing such mixtures at the farm or at home, care should be taken to avoid the uneven mixing of incompatible fertilizers, or the mixing which leads to a loss of some of the fertilizing nutrients in the form of gas, converts soluble nutrients into insoluble ones or induces caking. It is unwise to mix the following:

1. Ammonium sulphate, ammonium chloride, other ammoniacal fertilizers and nitrogenous organic manures with lime.
2. Sodium nitrate or potassium nitrate with superphosphate.

3. Nitrochalk with superphosphate or lime.
4. Ammonium sulphate-nitrate with lime.
5. Urea with superphosphate.
6. Superphosphate with lime or calcium carbonate or wood ashes.

Ammonium nitrate is an explosive chemical and, therefore, dangerous should be mixing at home. Mixtures containing other nitrates should be made in only such quantities as are to be used immediately, because they absorb water quickly and are not suitable for storage. Bone meal, sulphate of potash and muriate of potash can be mixed with all fertilizers. For further guidance or information on the method of mixing fertilizers, the State Agricultural Chemist or the nearest officer of the Government Agricultural Department should be contacted.

Mode and Time of Application of Fertilizers and Manures

Bulky organic manures should be applied well ahead of sowing, so that the preliminary decomposition takes place before the seeds germinate. Failing presowing application, they may be applied any time after the seedlings have established themselves. They are best applied in the powedered form. A sufficient supply of moisture in the soil is essential for their rapid decomposition. In the case of inorganic fertilizers, potassic and phosphatic fertilizers are best applied just before sowing or transplanting. Nitrogenous fertilizers may be applied either at planting and partly later. Split application is particularly desirable for nitrogen when applied to irrigated crops or to crops in heavy-rainfall areas.

Fertilizers applied before sowing should be broadcast uniformly and harrowed in. In the case of fertilizers containing soluble phosphate, the desirability of applying them in 2.5 to 5 cm wide bands on each side of the row of seeds at a depth of 10 to 15 cm with a drill has already been pointed out. This operation reduces the fixation of soluble phosphate in the soil. It is also a good practice to mix superphosphate with farmyard manure at 18 to 22 kg to a tonne before applying the organic manure, particularly of dairy farms land. Sulphate of ammonia used as a top-dressing should not be applied when plant leaves are wet. In the case of irrigated crops, the application of fertilizer should invariably be followed by a watering. In the case of fruit-trees, the fertilizer should be applied to the soil under the crown, a few metres away from the trunk. The area of application should be progressively extended as the trees grow bigger.

In advanced countries, fertilizers are usually applied with the help of machinery of diverse kinds and sometimes with an aeroplane or a helicopter. Combined-planters and fertilizer-distributors are employed when row crops are fertilized at sowing time.

Diagnosing the Fertilizer needs of Soils

There are four methods of determining the fertilizer requirements of soil:

a. Field experiments,

b. Pot tests,

c. Biological tests, and

d. Chemical test.

Field experiments contribute the more relaible method, but being time-consuming and expensive, they are conducted mainly by the research farms and research organisations. Farmers wishing to use field experiments as a Valid means of determining the fertility status of their soils should seek the advice of the state agronomist. Improperly conducted field experiments not only mean economic fertilizer practices.

Pot experiments permit test witha alarge number of manurial treatments within a limited space and in a relatively short time. However, as the conditions of such tests are different from those in the field, the results are not always directly applicable to large-scale farming.

Biological tests involve the growth of seedlings or of lower forms of plants, such as fungi and bacteria, under specified conditions and the study of their relative growth or the content of needed nutrients. A peridic testing of plant tissues for nitrates and other nutrients indicates the changing needs of crops for different food elements. But these are slow and costly processes and hence not always practicable.

The chemical analysis of soils or of plants growing on them constituents the modern method of determining the fertility status of a soil. Such analysis give information on the relative abundance or scarcity of the different nutrients required by crops from the soil, but they give no indication regarding the exact quantity, of fertilizer that may be applied to make good the deficiency. The dependability of this, method can, however, be increased a great deal by co-ordinating its results with those obtained from field experiments.

Facilities for rapid soil-testing have been made available in almost all states and can be availed of. At the same time, a very large number of fertilizer experiments with different crops on a variety of soils are conducted annually in all parts of the country under a comprehensive scheme. The results of these field experiments, when calibrated against those of rapid soil tests, will make the latter purely dependable. This will then be a valuable aid in the hands of extension workers in furnishing advice to farmers regarding fertilizer practices.

It is also sometimes possible to obtain a clue to the nutrient deficiences of soil with the help of deficiency symptoms in plants, as described already. However, a correct deagnosis of deficiency symptoms needs extensive experience. Furthermore, such symptoms in plants appear long after the actual occurance of nutrient deficiency in the soil. Therefore, such soil deficiencies must be diagnosed and remedied much earlier by adopting other means.

How much Fertilizer to Apply

In a vast country, such as India, possessing a wide variety of climatic and soil conditions, no definite quantity of any fertilizer can be prescribed as an optimum dose even for one and the same crop for all regions. each state in the country has conducted fertilizer investigations for the last 50 years and accumulated information on the manurial requirements of principal crops under local conditions. Therefore, for specific fertilizer practices in relation to particular crops and soils, the state agricultural department should always be consulted.

Considerations Governing the use of Fertilizers

It may be stressed that in order to secure the maximum response to fertilizers, the crop should be irrigated immediately, following their application and at suitable intervals thereafter. The response of a crop under arid and semi-arid conditions is usually uncertain and relatively small. It is, therefore, advantageous to restrict the use of chemical fertilizers mainly ti irrigated lands and areas of assured rainfall.

It must also be borne in mind that the maximum profit from the use of fertilizers depends on many factors, such as the nature of soil, the kinds of crop grown, the climate (in relation to soil, the kinds of crops growth), the prices of fertilizers, the market price of the agricultural produce and so no. All these factors must be given due consideration to secure the most economic results. Changes in one or more of them are bound to influence the economic results. changes in one or more of them are bound to influence the economics of fertilizer use. The use of fertilizers is also subject to the 'law of diminishing returns'. This means that the rate of increase in the crop yield decreases after a certain point is reached regarding the quantity of fertilizer used, and consequently the value of the additional yield finally becomes less than the cost of the fertilizer. Usually, the smaller applications of fertilizers produce greater percentage increase of yield than larger applications.

It should also be exphasized that an adequate supply of nutrients is only one of the factors that determine crop yield and that the application of fertilizers is not the sole means of making good the nutrient deficiencies in soils and plants. Equal attention must be paid to the other soil and crop-management practices to ensure good tilt, proper drainage, the required soil reaction, soil conversion, good land use, a suitable crop rotation, adequate organic matter in the soil and satisfactory soil micro-organism activity. Each one of these plays a vital role in determining the eventual productioin. the neglecting of one or more or these factors leads to a reduction in yield and created the need for still heavier manuring. Finally, manuring should not only be balanced in itself, but should also be designed to supplement the good effects of proper land use and beneficial soil management.

Chapter 2

Reproduction of Plants

The various modes in which plants reproduce their species may be conveniently classified into two groups, namely, *vegetative propagation* and *true reproduction,* the distinction between them being roughly this, that whereas in the former the production of the new individual may be effected by the most various parts of the body, in the latter it is always effected by means of a specialised reproductive cell.

PLANTS

Vegetative Propagation

The simplest case of vegetative multiplication is afforded by unicellular plants. When the cell which constitutes the body of the plant has attained its limit of size it gives rise to two either by division or gemmation; the two cells then grow, and at the same time become separated from each other, so that eventually two new distinct individuals are produced, each of which precisely resembles the original organism. A good example of this is to be found in the germination of the yeast plant. This mode of multiplication is simply the result of the ordinary processes of growth. All plant-cells grow and divide at some time or other of their life; but whereas in multicellular plant the products of division remain coherent, and add to the number of the cells of which the plant consists, in a unicellular plant they separate and constitute new individuals. In more highly organized plants vegetative propagation may be effected by the separation of the different parts of the body from each other, each such part developing the missing members and thus constituting a new individual. This takes place spontaneously in rhizomatous plants, in which the main stem gradually dies away from behind forwards; the lateral branches thus become isolated and constitute new individuals.

The remarkable regenerative capacity of plant-members is largely made use of for the artificial propagation of plants. A branch removed from a parent-plant will, under appropriate conditions, develop roots, and so constitute a new plant; this is the theory of propagation by "cuttings." A portion of a root will similarly develop one or more shoots, and thus give rise to a new plant. An isolated leaf will, in many cases, produce a shoot and a root, that is, a new plant; it is in this way that new begonias, for instance, are propagated. The production of plants from leaves occurs also in nature, as, for instance, in certain so-called "viviparous" plants, of which *Bryophyllum calycinum* (Crassulaceae) and many ferns are examples. But it is in the mosses, of all plants, that the capacity for vegetative propagation is most widely diffused. Any part of a moss, whether it be the stem, the leaves, the rhizoids, or the sporogonium, is capable, under appropriate conditions, of giving rise to filamentous protonema, on which new moss-plants are then developed as lateral buds.

In a large number of plants provision is made for vegetative propagation by the development of more or less highly specialized organs. In lichens, for instance, there are the *soredia,* which are minute buds of the thallus containing both algal and fungal elements; these are set free on the surface in large numbers, and each grows into a thallus. In the Characeae there are the *bulbils* or "starch-stars" of *Chara stelligera,* which are underground nodes, and the branches with naked base and the proembryonic branches found by Pringsheim on old nodes of *Chara fragilis.* In the mosses small tuberous bulbils frequently occur, on the rhizoids, and in many instances stalked fusiform or lenticular multicellular bodies containing chlorophyll, termed *gemmae,* are produced on the shoots, either in the axils sof the leaves or in special receptacles at the summit of the stem. Gemmae of this kind are produced in vast numbers in *Marchantia* and *Lunularia* among the liverworts. Similar gemmae are also produced by the prothallia of ferns. In some ferns the buds borne on the leaves or in their axils become swollen and filled with nutritive materials, constituting *bulbils* which fall off and give rise to new plants. This conversion of buds into bulbils, which subserve vegetative multiplication, occurs also occasionally among Phanerogams, as for instance in *Lilium bulbiferum*, species of *Poa*, *Polygonum viviparum*, &c. But many other adaptations of the same kind - occur among Phanerogams. Bulbous plants, for instance, produce each year at least one bulb or corm from which a new plant is produced in the succeeding year. In the potato, tubers are developed from subterranean shoots, each of which in the following year gives rise to a new individual. In the dahlia, *Thladiantha dubia*, &c., tuberous swellings are found on the roots, from each of which a new individual may spring.

True Reproduction

This is effected by cells formed by the proper reproductive organs. These cells are of two principal kinds. There are, first, those cells each of which is capable of developing by itself into a new organism: these are the *asexual* reproductive cells, known generally as *spores*. Secondly, there are the cells which are incapable of independent germination; it is not until these cells have fused together in pairs that a new organism can be developed: these are the *sexual* reproductive cells or *gametes*. In some exceptional cases the normal mode of reproduction, sexual or asexual, does not take place: instead, the new organism is developed vegetatively from the parent. When sexual reproduction is suppressed the case is one of *apogamy;* when asexual reproduction by spores is suppressed the case is one of *apospory*. (Apogamy and apospory are discussed below in the section on *Abnormalities of Reproduction.) Asexual Reproduction.* - *Reproduction* by means of some kind of spore (using the term in its widest sense, so as to include all asexually produced reproductive cells) is common to nearly all families of plants; it is wanting in certain

Algae (Conjugatae, Fucaceae, Characeae), and in certain fungi. The structure of a spore is essentially this: it consists of a nucleated mass of protoplasm, enclosing starch or oil as reserve nutritive material, usually invested by a cell-wall. In those cases in which the spore is capable of germinating immediately on its development the cell-wall is a single delicate membrane consisting of cellulose; but in those cases in which the spore may or must pass through a period of quiescence before germination the wall becomes thickened and may consist of two layers, an inner, the *endospore,* which is delicate and consists of cellulose, and an outer, the *exospore,* which is thick and rigid, frequently darkly coloured and beset externally with spines or bosses, and which consists of cutin. In some few cases among the fungi, multicellular or septate spores are produced; these approximate somewhat to the gemmae mentioned above as highly specialized organs for vegetative propagation. In some cases, particularly among the algae, and also in some fungi (Peronosporeae, Saprolegnieae, Chytridiaceae, and the Myxomycetes), spores are produced which are usually destitute of any cell-wall, and are further peculiar in that they are motile, and are therefore termed *zoospores;* they move sometimes in an amoeboid manner by the protrusion of pseudopodia, but more frequently they are provided with one, two, or many delicate vibratile protoplasmic filaments, termed *cilia,* by the lashing of which the spore is propelled through the water. The zoospore eventually comes to rest, withdraws its cilia, surrounds itself with a cell-wall, and then germinates.

In the simplest case a single spore is developed from the cell of the unicellular plant, the protoplasm of which surrounds itself with the characteristic thick wall. This occurs only in plants of low organization such as the Schizophyta.

In other cases the contents of the cell undergo division, each portion of the protoplasm constituting a spore. Examples of this are afforded, among unicellular plants, by yeast and the Protococcaceae; and in multicellular plants by the Pandorineae, Confervaceae, Ulvaceae, &c., where any cell of the body may produce spores.

In such cases the spore-producing cell may be regarded as a rudimentary reproductive organ of the nature of a *sporangium.* In more highly organized plants special organs are differentiated for the production of spores. In the majority of cases the special organ is a sporangium, that is, a capsule in the interior of which the spores are developed; but in many fungi the spores are formed by abstriction from an organ termed a *sporophore.* In the Thallophyta the sporangium is commonly a single cell. In the Bryophyta it is a multicellular capsule. In the Pteridophyta the sporangium is multicellular, but simple in structure, and this is true also of the Phanerogams.

It is important to note that in all the Bryophyta and in some of the Pteridophyta (most of the Filicinae, all existing Equisetinae, and the Lycopodiaceae and Psilotaceae) there is but one kind of sporangium and spore, the plants being *homosporous* or isosporous, whereas the rest of the Pteridophyta (Hydropterideae, Selaginellaceae) and the Phanerogams are *heterosporous,* having sporangia of two kinds; some produce one or a few large spores (*megaspores*), and are hence termed *megasporangia,* while others give rise to a larger number of small spores (*microspores*) and are hence termed *microsporangia.* In the Phanerogams the two kinds of sporangia have received special names: the megasporangium, which produces as a rule only one mature spore (*embryo-sac*), is termed the *ovule;* the microsporangiurn, which produces a large number of microspores (*pollen-grains*), is termed the *pollen-sac.* The development of spores, except in the simpler Thallophyta, is more or less restricted to definite parts of the body. Thus in the Red Algae (Florideae) there are the organs known as *stichidia, nemathecia.* In the fungi the number and variety of such organs is very great; they may be described generally as simple and compound *sporophores:* but for a description the article Fungi should be consulted. In the higher plants the organs are less various. In the Bryophyta the production of spores is restricted to the sporogonium. In the vascular plants (Pteridophyta, Phanerogams) the development of sporangia, speaking generally, is confined to the leaves. In most ferns the sporangiferous leaves (*sporophylls*) do not differ in appearance from the foliage leaves; but in other Pteridophyta (Equisetaceae, Marsiliaceae, some species of *Lycopodium* and *Selaginella*) they present considerable adaptation, and notably in the Phanerogams. In the Phanerogams the specialization is so great that the sporophylls have received special names; those which bear the microsporangia (pollen-sacs) are termed the *stamens,* and those which bear the megasporangia (ovules) are termed the *carpels*. The sporophylls are usually aggregated together on a short stem, forming a shoot that constitutes a flower.

Many terms are employed to indicate the nature of the various kinds of spores, especially among the fungi, but the endless varieties of asexual reproductive cells may be grouped under two heads - (r) *Gonidia,* (2) *Spores proper*. The distinction between these two kinds of asexual reproductive cells is as follows.

The *gonidium* is a reproductive cell that gives rise, on germination, to an organism resembling the parent. For instance, among the algae, the "zoospore" of *Vaucheria* develops into a Vaucheria-plant. There is thus a close connexion between vegetative multiplication and multiplication by means of gonida. The production of gonida is entirely limited to the Thallophyta, and is especially marked in the fungi, though the nature of all the many kinds of reproductive cells formed in this group has not yet been fully investigated. It is, however, wanting in certain algae and fungi.

The *spore proper* is a reproductive cell that as a rule gives rise, on germination, to an organism unlike that which produced it. For instance, the spore of a fern when it germinates gives rise, not to a fern-plant, but to a prothallium. The apparent exceptions to this rule occur only among the Thallophyta, and are explained below in the section on *Life-history*. The true spore is developed, usually in a sporangium, after a process of division which presents certain features that call for special notice.

Observation of the process of division of the nucleus (*karyokinesis*) in plants generally has shown that the linin-reticulum of the resting nucleus breaks up into a definite number of segments, the *chromosomes,* each of which bears a series of minute bodies, the *chromatin-disks* or *chromomeres,* consisting largely of a substance termed *chromatin.* In the ordinary *homotype* divisions of the nuclei the characteristic number of chromosomes is always observable: but when the spore-mother-cells are being formed the number of chromosomes is reduced to one-half. This, if the number of chromosomes of the parent plant be expressed as $2x$ the number in the spore will be x. To take a concrete case: it has been observed by Guignard and others that in the early divisions taking place in the developing anther and ovule of the lily the number of chromosomes is 24; whereas in the later divisions which give rise to the pollen-mother-cells in the one case and to the mothercell of the embryo-sac in the other, the number of chromosomes is only 12.

Thus the development of a spore is always preceded by a *reducing-* or *heterotypedivision,* a process now more generally termed *meiosis* (Farmer). The reduced number of chromosomes in the nucleus of the spore-mother-cell persists in the spore, and in all the cells of the organism to which the spore may give rise. It should be explained that cells, to which the name "spore" has also been applied, are formed as the result of a sexual act: such are *zygospores, oospores,* and some *carpospores.* But these cells differ from spores proper not only in their mode of origin but also in that their nuclei contain the full double number (2x) of chromosomes; hence they may be distinguished as *diplospores. Sexual Reproduction. - Sexual* reproduction involves the development of sexual organs (*gametangia*) and sexual cells (*gametes*). When the organism is unicellular, as in the lower Green Algae (e.g. Protococcaceae, Conjugatae), the cell becomes a sexual organ and its whole protoplasm gives rise to one or more sexual cells: in the higher forms certain parts of the body are specialized as sexual organs. In many of the lower plants the organs present no external distinction of sex it is impossible to distinguish between the male and female organs, although it cannot be doubted that the essential physiological difference exists; consequently the organs are merely described as gametangia. The gap between these plants and those with differentiated sexual organs is, however, bridged over by intermediate forms, as explained in the article Algae.

When the sexual organs are more or less obviously differentiated into male and female, they present considerable variety of form in different groups of plants, and accordingly bear different names. Thus the male organ is a *pollinodium* in most of the fungi, a *spermogonium* in others (certain Ascomycetes, Uredineae); in all other plants it is an *antheridium*. Similarly the female organ is an *oogonium* in various Thallophyta (Green and Brown Algae: Oomycetous Fungi); a *procarp* in the Red Algae; an *archicarp* in certain Ascomycetous Fungi and in the Uredineae; an *archegonium* in all the higher plants.

It is generally the case that the protoplasm of the sexual organ is differentiated into one or more sexual cells. Thus the gametangium usually gives rise to cells which, as they are externally similar, are termed *isogametes* or simply *gametes*. Certain forms of the male organ, the spermogonium and the antheridium, give rise to male cells which are termed *spermatic* when they are nonciliate, *spermatozoids* when they are ciliated and free-swimming. Again, the female organs termed oogonia and archegonia produce one or more female cells called *oospheres*. But there are important exceptions to this rule. Thus the protoplasm is not differentiated into cells in the gametangium of the Mucorinae; in the male organ (pollinodium), of fungi generally; and in the female organ (procarp) of the Red Algae and (archicarp) of the Ascomycetes and Uredineae.

The immediate product of the fusion of cells, or of undifferentiated protoplasm, derived from sexual organs of opposite sex may be generally termed the *zygote;* but it is not always of the same kind. Thus when two isogametes, or the undifferentiated contents of two gametangia, fuse together, the process is designated *conjugation*, and the product is usually a single cell termed *zygospore*. When an oosphere fuses with a male cell, or with the undifferentiated contents of a male organ, the process is *fertilization*, and the product is a single cell termed *oospore*. When, finally, a female organ with undifferentiated contents receives a male cell, the process again is fertilization; here the product is not a single cell, but a fructification termed *cystocarp* (Red Algae), or *ascocarp* (Ascomycetes) or *aecidium* (Uredineae), containing many spores (*carpospores)*. As a consequence of the diversity in the sexual organs and cells, in the details of the sexual act, and in the product of it, several modes of the sexual process have to be distinguished, which may be conveniently summarized as follows:

1. *Isogamy:* the sexual process consists in the fusion of either two similar sexual cells (*isogametes*), or two similar sexual organs (*gametangia)* : it is termed conjugation, and the product is a *zygospore*. Its varieties are
   a. Gametes ciliated and free-swimming (*planogametes*), set free into the water where they meet and fuse: lower Green Algae (Protococcaceae, Pandorineae, most Siphonaceae and Confervaceae); some Brown Algae (Phaeosporeae):
   b. Gametangia fuse in pairs, and a gamete is differentiated in each: the gametes of each pair fuse, but are not set free and are not ciliated (the Conjugate Green Algae): or, no gametes are differentiated, the undifferentiated contents of the gametangia fusing (Mucorinae among the Fungi).
2. *Oogamy:* male and female organs distinct: the protoplasm, of the female organ is differentiated into one or (rarely) more oospheres which usually remain enclosed in the female organ: the contents of the male organ are usually differentiated into one or more male cells: the process is fertilization, the product is

an *oospore.* (A) The sexual organs are unicellular (or coenocytic as in certain Siphonaceous Green Algae and in the Oorycetous Fungi); the female organ is an *oogonium.*

a. The male organ is an *antheridium* giving rise to one or more free-swimming ciliated spermatozoids:
   - The oogonium contains a single oosphere which is fertilized *in situ:* higher Green Algae (*Volvox, Vaucheria, Oedogonium, Coleochaete,* Characeae); some Brown Algae (*Tilopteris);* among the Fungi, *Monoblepharis,* the only fungus known to have spermatozoids:
   - The oogonium produces a single oosphere which is extruded and is fertilized in the water: *Dictyota* and some Fucaceae (Brown Algae):
   - The oogonium contains several oospheres which are fertilized *in situ: Sphaeroplea* (Siphonaceous Green Alga):
   - The oogonium produces more than one oosphere (2-8) which are extruded and are fertilized in the water: certain

   Brown Algae (*Pelvetia, Ascophyllum, Fucus):* The male organ is a *pollinodium* which applies itself closely to the oogonium: the amorphous male cell is not ciliated and is not set free:
   - The oogonium contains a single oosphere which is fertilized *in situ:* Peronosporaceae (Oomycetes):
   - The oogonium contains several oospheres; Saprolegniaceae: but it is debated whether or not fertilization actually takes place.

b. The male and female organs are (as a rule) multicellular; the male organ is an *antheridium,* the female an *archegonium:* the archegonium always contains a single oosphere which is fertilized *in situ.*

- The male cell is a free-swimming ciliated spermatozoid: the antheridium produces more than one (usually very many) spermatozoids, each of which is developed in a single cell: all Bryophyta and Pteridophyta: the only Phanerogams in which spermatozoids have been observed are the gymnospermous species *Ginkgo biloba, Cycas revoluta, Zamia integrifolia.*

(13) The male cell is amorphous and passes directly from the pollen-tube into the oosphere (*siphonogamy)* : all Phanerogams except the species just mentioned.

It must be explained that in the angiospermous Phanerogams, the male and female organs are so reduced that each is represented by only a single cell: the male, by the *generative cell,* formed in the pollen-grain, which usually divides into two male cells: the female, by the oosphere. The gradual reduction can be traced through the Gymnosperms.

Attention may here be drawn to the fact that, in several cases, the second male cell has been seen to enter the embryo-sac from the pollen-tube, and its nucleus to fuse with the definitive nucleus (endosperm-nucleus) or with one of the polar nuclei.

3. *Carpogamy:* the sexual organs are (as a rule) differentiated into male and female: the protoplasm of the unicellular or multicellular female organ (archicarp, procarp) is never differentiated into an oosphere: in many cases definite male cells, *spermatia,* are produced and are set free, but they are not ciliated, and frequently have a cell-wall: the process is *fertilization:* the product is a fructification derived essentially from the female organ con taining several (sometimes very many) spores (*carpospores):* characteristic of the Red Algae and of the Ascomycetous Fungi.

a. There are definite male cells (*spermatia):* (a) The female organ is a *procarp,* consisting of an elongated, closed, receptive filament, the *trichogyne,* and of a basal fertile portion, the *carpogonium:* on fertilization the latter grows and gives rise directly or indirectly to a *cysto-* carp: the spermatia are each formed in a unicellular antheridium and have no cell-wall at first: they fuse with the tip of the trichogyne: Red Algae (Rhodophyceae or Florideae): (0) The female organ (*archicarp*) resembles the preceding: in fertilization the fertile portion (*ascogonium*) develops into an *ascocarp* containing one or more asci (sporangia) each containing usually eight *ascospores:* the spermatia are formed by abstriction from the filaments (*sterigmata)* lining special receptacles, the *spermogonia,* which are the male organs: certain Ascomycetous Fungi (e.g. Laboulbeniaceae, some Lichen-Fungi, *Polystigma).*

b. There are no definite male cells: the more or less distinct male and female organs come into contact, and their undifferentiated contents fuse: the product is an *ascocarp:* (a) The male and female organs are obviously different: the female organ is an *ascogonium,* the male a *pollinodium: e.g. Pyronema, Sphaerotheca* (Ascomycetes): (R) The male and female organs are quite similar: *e.g. Eremascus, Dipodascus* (Ascomycetes).

It may be explained that carpogamy is the expression of sexual degeneration. In the cases last mentioned, when the sexual organs are quite similar, they have reverted to the condition of gametangia. Still further reduction is observable in other Ascomycetes in which one of the sexual organs, presumably the male, is either much reduced or is altogether wanting. Again in the rusts (Uredineae), there are spermatia, but they are functionless. In the highest Fungi, the Autobasidiomycetes, no sexual organs have been discovered.

DETAILS OF THE SEXUAL ACT

It has been already stated that the sexual act consists in the fusion of two masses of protoplasm, commonly cells, derived from two organs of opposite sex: but this is only the first stage in the process. The second stage is the fusion of the nuclei, which usually follows quickly upon the fusion of the cells; but nuclear fusion may be postponed so that the two sexual nuclei may be observed in the zygote, as "conjugate" nuclei, and even in the cells of the organism developed from the zygote (e.g. Uredineae).

The result of nuclear fusion is that the nucleus of the zygote contains the double number of chromosomes - that is, if the number of chromosomes in each of the fusing sexual nuclei be x, the number in the nucleus of the zygote will be 2X. Moreover, this double number persists in all the cells of the organism developed from the zygote, until it is reduced to one-half by meiosis preceding either the development of the spores, or, less commonly, the development of the sexual cells. But there is yet a third stage, which consists in the temporary fusion of the chromosomes belonging to the two sexual nuclei. This always takes place as a preliminary to meiosis; it may be in thc germinating zygote, or after many generations of cells have been formed from it.

At the onset of meiosis the (2x) chromosomes are seen to be double, one of each pair having been derived from the male and the female cell respectively: the chromosomes of each pair then fuse so that their chromomeres unite along their length, constituting the *pseudo-chromosomes.* The paired chromosomes separate and eventually go to form the two daughter-nuclei, one to each, which thus have half (x) the original number of chromosomes. The daughter-nuclei at once divide homotypically, retaining the reduced (x) number of chromosomes to form the four nuclei of a tetrad of spores (more rarely, *e.g. Fucus,* of sexual cells).

Life-history

It will have been gathered from the foregoing sections that plants generally are capable of both sexual and asexual reproduction; and, further, that in different stages of their life-history they possess the diploid (2x) number of chromosomes in their nuclei, or the haploid (x) number. It may be at once stated that, in all plants in which sexual reproduction and true meiotic spore-formation exist, these two modes of reproduction are restricted to distinct forms of the plant; the sexual form bears only the sexual organs and is haploid; the asexual form only produces spores and is diploid. Hence all such plants are to this extent *polymorphic - that* is, the plant assumes these two forms in the course of its life-history. When, as in many Thallophyta, one or other of these forms can reproduce itself by means of gonidia, additional forms may be introduced into the lifehistory, which becomes the more complicated the more pronounced the polymorphism.

The most straight forward life-histories are those presented by the Bryophyta and the Pteridophyta, where there are but the two forms, the sexual and the asexual. In the life-history of a moss, the plant itself bears only sexual organs: it is the sexual form, and is distinguished as the *gametophyte.* The zygote (oospore) formed in the sexual act develops into an organism, the *sporogonium,* which is entirely asexual, producing only spores: it is distinguished as the *sporophyte.* When these spores germinate, they give rise to moss-plants. Thus the two forms, the sexual and the asexual, regularly alternate with each other - that is, the life-history presents that simple form of polymorphism which is known as *alternation of generations.* Similarly, in the life-history of a fern, there is a regular alternation of a sporophyte, which is the fern-plant itself, with a gametophyte, which is the fern-prothallium.

It is pointed out in the preceding section that, as the result of the sexual act, the nucleus of the zygote contains twice as many chromosomes as those of the fusing sexual cells. This 2X number of chromosomes persists throughout all the cellgenerations derived from the zygote, that is, in the cells constituting the sporophyte, up to the time that it begins to produce spores, when meiosis takes place. Again, the cellgenerations derived from the spore, that is, the cells constituting the gametophyte, all have the reduced *x* number of chromosomes in their nuclei up to the sexual act. Hence the sporophyte may also be designated the *diplophyte* and the gametophyte the *haplophyte* (Strasburger): in other words, the sporophyte is the *pre-meiotic*, the gametophyte the post-meiotic generation. Twice in its life-history the plant is represented by a single cell: by the spore and by the zygote. The turningpoints in the life-history, the transitions from the one generation to the other, are (r) meiosis, (2) the sexual act.

The course of the life-history in Phanerogams and in those Thallophyta which have been adequately investigated is essentially the same as that of the Bryophyta and of the Pteridophyta as described above, though it is less easy to trace on account of the peculiar relation of the two generations to each other in the Phanerogams and on account of various irregularities that present themselves in the Thallophyta.

In the Phanerogams, as in the Pteridophyta, the preponderating generation is the sporophyte, the plant itself. Inasmuch as they are heterosporous, the gametophyte is represented by a male and a female organism or prothallium, both rudimentary. The male prothallium consists of the few cells formed by the germinating pollen-grain (microspore); and though it is quite independent, since the microspores are shed, it grows parasitically in the tissues upon which the microspore has been deposited in pollination. The female prothallium may consist of many cells with well-developed archegonia, as in the Gymnosperms, or of only a few cells with the female organ reduced to the oosphere, as in the Angiosperms. In either case it is the product of the germination of a megaspore (embryo-sac) which is not shed from its sporangium (ovule): hence it never becomes an independent plant, and was long regarded as merely a part of the sporophyte until its true nature was ascertained, chiefly by the researches of Hofmeister, who first explained the alternation of generations in plants. This intimate and persistent connexion between the two generations affords the explanation of the characteristic features of the Phanerogams, the seed and the flower.

The ovule containing the embryo-sac, which eventually contains the embryo, persists as the seed - a structure that is distinctive of Phanerogams, which have, in fact, on this account been also termed Spermatophyta. With regard to the flower, it has been already mentioned that it is, like the cone of an Equisetum or a Lyco podium, a shoot adapted to the production of spores. But it is something more than this: for whereas in Equisetum or Lycopodium the function of the cone comes to an end when the spores are shed, the flower of the Phanerogam has still various functions to perform after the maturation of the spores. It is the seat of the process of *pollination - that* is, the bringing of the pollen-grain by one of various agencies into such a position that a part (the pollen-tube) of the male prothallium developed from it may reach and fertilize the oosphere in the embryo-sac. Thus the flower of Phanerogams is a reproductive shoot adapted not only for spore-production, but also for pollination, for fertilization, and for the consequences of fertilization, the production of seed and fruit. However, in spite of these complications, it is possible to determine accurately the limits of the two generations by the observation of the nuclei. The meiosis preceding the formation of the spores marks the beginning of the (haploid) gametophyte, male and female; and the sexual act marks that of the (diploid) sporophyte.

The difficult task of elucidating the life-histories of the Thallophyta has been successfully performed in certain cases by the application of the method of chromosome-counting,. with the result that alternation of generations has been found to be of general occurrence. To begin with the Algae. In. the Dictyotaceae (Brown Algae) there are two very similar forms in the life-history, the one bearing asexual reproductive organs (tetrasporangia), the other bearing sexual organs (oogonia. and antheridia). It has been shown (Lloyd Williams) that the former is undoubtedly the sporophyte and the latter the gametophyte, since the nuclei of the former contain 3 2 chromosomes, and those of the latter 16. Meiosis takes place in the mother-cell of the tetraspores, which, on germination, give rise to the sexual form. Quite a different life-history has been traced in *Fucus,* another Brown Alga. Here no spores are produced: there is but one form in the life-history, the *Fucus* plant, which bears sexual organs and has, on that account, been regarded as a gametophyte. The investigation of the nuclei has, however, shown (Farmer) that the Fucus-plant is actually diploid, that it is, in fact, a sporophyte; but since there is no spore-formation, meiosis immediately precedes the development of the sexual cells, which alone represent the gametophyte.

Similarly, two types of life-history have been discovered in the Red Algae. In *Polysiphonia violacea,* a species in which. the tetraspores and the sexual organs are borne by similar but. distinct individuals, it has been ascertained (Yamanouchi) that, as in *Dictyota,* meiosis takes place in the mother-cell of the tetraspores, so that the nuclei of these spores, as also those of the sexual plants to which they give rise, contain 20 chromosomes: and further, that the nuclei of the carpospores (diplospores) produced in the cystocarp as the result of fertilization,. contain 40 chromosomes, as do also those of the asexual plant. to which the carpospores give rise. Hence the sporophyte is represented by the cystocarp and the resulting tetrasporangiate plants: the gametophyte, by the sexual plants. Though it is the rule in the Red Algae that the tetrasporangia and the sexual organs are borne on distinct individuals, yet cases are known in which both kinds of reproductive organs arse borne upon the same plant; and to those the above conclusions obviously cannot apply. They have yet to be investigated.

The second type of life-history has been traced in *Nemalion.* Here there is no tetrasporangiate form, consequently meiosis. takes place at a different stage in the life-history. It has. been observed (Wolfe) that the nuclei of the sexual plant contain 8 chromosomes; those of the gonimoblast-filaments of the developing cystocarp contain r6, whilst those of the carpospores contain 8: hence meiosis takes place in the carposporangia. Here the plant is the gametophyte; the sporophyte is only represented by the cystocarp. The carpospores here are true spores (haplospores).

Among the Green Algae, *Çoleochaete* is the only form that has been fully investigated (Allen). Here meiosis takes place in the germinating oospore: consequently the plant is the gametophyte, and the sporophyte is represented only by the oospore, so that the life-history resembles that of *Nemalion*. It is probable that this conclusion is generally true of the whole group; at any rate of those forms (Desmids, *spirogyra, Oedogonium, Chara*) which have been more or less investigated.

Turning to the Fungi, somewhat similar results have been obtained in the few forms that have been studied from this point of view. In the sexual Ascomycetes it appears (Harper) that meiosis takes place in the ascocarp just before the development of the spores, so that the life-history essentially resembles that of *Nemalion*. Again, in certain Uredineae, having an aecidium-stage and a teleutospore-stage, which is apparently a sexual process has been observed (Blackman, Christman) which is described in the section on *Abnormalities of Reproduction*, and the life-history is as follows. The sexual act having taken place, a row of aecidiospores is developed in the aecidium, each of which contains two conjugate nuclei derived from the sexual nuclei. The mycelium developed from the aecidiospore, as well as the uredospores and the teleutospores that it bears, shows two conjugate nuclei. When, however, the teleutospore is about to germinate, the two nuclei fuse (thus completing the sexual act) and meiosis takes place. As a result the promycelium developed from the teleutospore, and the sporidia that it produces, are uninucleate: so are also the mycelium developed from the sporidium, and the female organs (archicarps) borne upon it. Hence the limits of the sporophyte are the aecidiospore and the teleutospore: those of the gametophyte, the teleutospore and the aecidiospore.

Similar observations have been made upon other Uredineae with a more contracted life-history. *Phragmidium Potentillaecanadensis* is a rust that has no aecidium-stage: consequently the primary uredospores are borne by the mycelium produced on infection of the host by a sporidium. It has been observed (Christman) that the sporogenous hyphae fuse in pairs, suggesting a sexual act; then the primary uredospores are developed in rows from the fused pairs of hyphae which thus behave as sexual organs (archicarps), and each such uredospore contains two conjugate nuclei. Although the research has not been carried beyond this point, it may be inferred that in this case, as in the preceding, nuclear fusion and meiosis take place in the teleutospore. Here the sporophyte is represented by the uredo-form.

Finally, in some of the fungi in which no sexual organs have yet been discovered, this method of investigation has made it probable that some kind of sexual act takes place nevertheless. Thus in the Uredine *Puccinia malvacearum*, which has only teleutosporeand sporidium-stages, it has been observed (Blackman) that the formation of the teleutospores is preceded by a binucleate condition of the hyphae. The same idea is suggested by the binucleate basidia of the Basidiomycetes, which correspond to the teleutospores of the Uredineae.

The life-histories sketched in the preceding paragraphs show that one of the complexities met with in the Thallophyta is that meiosis does not always take place at the same point in the life-history. In the higher plants the incidence of meiosis is generally, though not absolutely, constant: it may be stated as a rule that in the Bryophyta, Pteridophyta and Phanerogams it takes place in the spore-mother-cells. In the Thallophyta this rule does not hold. In some of them, it is true, meiosis immediately precedes, as in the higher plants, the formation of certain spores, the tetraspores (Dictyotaceae, *Polysiphonia*), the teleutospores (Uredineae): but in others it immediately precedes the development of the sexual organs (Fucaceae), or follows more or less directly upon the sexual act (Green Algae, *Nemalion*, Ascomycetes).

The life-history of most Thallophyta is further complicated by the capacity of the gametophyte of the sporophyte to reproduce themselves by cells termed gonidia, a capacity that is wholly lacking in the higher plants. The karyology of gonidia has not yet been sufficiently investigated: but when, as in the

Green Algae and the Oomycetous Fungi, the gonidia are developed by and reproduce the gametophyte, it may be inferred that they, like the gametophyte, are haploid. One case, at any rate, of the reproduction of the sporophyte by gonidia is fully known, that of the Uredineae just described, in which the uredoform, which is a phase of the sporophyte, is reproduced by the uredo-spores which are binucleate, that is diploid, and may be distinguished as *diplogonidia*. In any case the result is that whereas in the higher plants each of the alternating generations occurs but once in the life-history, in these Thallophyta the lifehistory may include a succession of gametophytic or of sporophytic forms, This is, in fact, a distinguishing feature of the group. The higher plants present a *regular* alternation of generations: whereas, in the Thallophyta, though they probably all present some kind of alternation of generations, yet it is *irregular* in the various ways and for the various reasons mentioned above.

Sufficient information has been given in the preceding pages to render possible the consideration of the *origin* of alternation of generations. To begin quite at the beginning, it may be assumed that the primitive form of reproduction was purely vegetative, merely division of the unicellular organism when it had attained the limits of its own growth. Following on this came reproduction by a gonidium: that is, the protoplasm of the cell, at the end of its vegetative life, became quiescent, surrounded itself with a proper wall, or was set free as a motile ciliated cell, having in some unexplained way become capable of originating a new course of life (*rejuvenescence*) on germination..

Then, as can be well traced in the Brown and Green Algae, these primitive reproductive cells (gonidia) began to fuse in pairs: in other words, they gradually became sexual. This stage can still be observed in some of these Algae (e.g. *Ulothrix*, *Rctocarpus*) where the zoospores (gonidia) may either germinate independently, or fuse in pairs to form a zygote. Gradually the sexuality of these cells became more pronounced: losing the capacity for independent germination, they acquired the external characters of more or less differentiated sexual cells, and the gametangia producing them developed into male and female sexual organs. But this advancing sexual differentiation did not necessarily deprive the plant of the primitive mode of propagation: the sexual organism still retained the faculty of reproduction by gonidia. The loss of this faculty only came with higher development: it is entirely wanting in some of the higher Thallophyta (e.g. Fucaceae, Characeae), and in all plants above them in the evolutionary series.

With the introduction of the sexual act, a new kind of reproductive cell made its appearance, the zygote. This cell, as. already explained, differs from other kinds of spores and from the sexual cells, in that its necleus is diploid; and with it the sporophyte (diplophyte) was introduced into the life-history. It has been mentioned that in some plants (e.g. Green Algae) the zygote is all that there is to represent the sporophyte, giving rise, or germination and after meiosis, to one or more spores. Passing to the Bryophyta, in the simpler forms (e.g. *Riccia*), the zygote develops into a multicellular capsule (sporogonium); and in the higher forms into a more elaborate sporogonium, producing many spores. In the Pteridophyta and the Phanerogams, the zygote gives rise to the highly developed sporophytic plant.

Thus the evolution of the sporophyte can be traced from the unicellular zygote, gradually increasing in bulk and in independence until it becomes the equal of the gametophyte (e.g. in *Dictyota* and *Polysiphonia*), and eventually far surpasses it (Pteridophyta, Phanerogams). Moreover, the increase in size was attended by the gradual limitation of spore-production to certain parts only, the rest of the tissues being vegetative, assuming the form of stems, leaves, &c. These facts have been formulated in the theory of "progressive sterilization" (Bower), which states that the sporophytic form of the higher plants has been evolved from the simple, entirely fertile, sporophyte of the lower, by the gradually increasing development of the sterile vegetative tissue at the expense of the sporogenous, accompanied by increase in total bulk and in morphological and histological differentiation.

In connexion with the study of the evolution of the sporophyte, the question arose as to its morphological significance; whether it is to be regarded as a modified form of the gametophyte, or as an altogether new form intercalated in the life-history: in other words, whether the alternation is "homologous" or "antithetic." In certain plants there is a succession of forms which are undoubtedly homologous: for instance, in *Coleochaete* where a succession of individuals without sexual organs is produced by zoospores (gonidia). The main fact that has been established is that the sporophyte, from the simple zygote of the Thallophyta to the spore-bearing plant of the Phanerogams, is characterized by its diploid nuclei; that it is a diplophyte, in contrast to the haplophytic gametophyte. Were these nuclear characters absolutely universal, there could be no question but that the sporophyte is an altogether new antithetic form, and not an homologous generation. But certain exceptions to the rule have been detected, which are described under *Abnormalities of Reproduction:* at present it will suffice to say that such things as a diploid gametophyte and a haploid sporophyte have been observed in certain ferns.

It can only be inferred that alternation of generations is not absolutely dependent upon the periodic halving in meiosis and the subsequent doubling by a sexual act, of the number of chromosomes in the nuclei, though the two sets of phenomena usually coincide. It must not, however, be overlooked that these exceptional cases occur in plants presenting an abnormal life-history: the fact remains that where there is both normal spore-formation with meiosis, and a subsequent sexual act, the haploid form is the gametophyte, the diploid the sporophyte. But the actual observation of a haploid sporophyte and of a diploid gametophyte makes it clear that however generally useful the nuclear characters may be in the distinction of sporophyte and gametophyte, they do not afford an absolute criterion, and therefore their value in determining homologies is debatable.

Abnormalities of Reproduction

In what may be regarded as the type of normal life-history, the transition from the one generation to the other is marked by definite processes: there is the meiotic development of spores by the sporophyte, and the sexual production of a zygote, or something analogous to it, by the gametophyte. But it has been mentioned in the preceding pages that the transition may, in certain cases, be effected in other ways, which may be regarded as abnormal, though they are constant enough in the plants in which they occur, in fact as manifestations of reproductive degeneration.

In the first place, the sporophyte may be developed either after an abnormal sexual act, or without any preceding sexual act at all, a condition known as *apogamy*. In the second, the gametophyte may be developed otherwise than from a post meiotic spore, a condition known as *apospory*. Apogamy. - The cases to be considered under this head may be arranged in two groups:

1. *Pseudapogarny: sexual act abnormal*: *The* following abnormalities have been observed:
   a. *Fusion of two female organs*: observed (Christman) in certain Uredineae (*Caeoma nitens, Phragmidium speciosum, Uromyces Caladii*) where adjacent archicarps fuse: male cells (spermatia) are present but functionless.
   b. *Fusion between nuclei of the same female organ*: observed in the ascogonium of certain Ascomycetes, *Humaria granulate* (Blackman), where there is no male organ; *Lachnea stercorea* (Fraser), where the male organ (pollinodium) is present but is apparently functionless.
   c. *Fusion of a female organ with an adjacent tissue-cell*: observed (Blackman) in the archicarp of some Uredineae (*Phragmidium violaceum, Uromyces Poae, Puccinia Poarum)* : male cells (spermatia) present but functionless.

d. There is no female organ: fusion takes place between two adjacent tissue-cells of the gametophyte; the sporophyte is developed from diploid cells thus produced, but there is no proper zygote as there is in *a, b* and *c* : observed (Farmer) in the prothallium of certain ferns (*Lastraea pseudo-mas,* var. *polydactyla) :* male organs (and sometimes female) present but functionless. Another such case is that of *Humaria rutilans* (Ascomvcete), in which nuclear fusion has been observed (Fraser) in hyphae of the hypothecium: the asci are developed from these hyphae, and in them meiosis takes place; there are no sexual organs.

2. *Eu-apogamy: no kind of sexual act* (a) The gametophyte is haploid:

a. The sporophyte is developed from the unfertilized oosphere: no such case of *true parthenogenesis* has yet been observed.

The sporophyte is developed vegetatively from the gametophyte and is haploid: observed in the prothallia of certain ferns, *Lastraea pseudo-mas,* var. *cristata-apospora(Farmer* and Digby), and *Nephrodium molle* (Yamanouchi).

b. *The gametophyte is diploid*: The sporophyte is developed from the diploid oosphere: observed in some Pteridophyta, viz. certain ferns (Farmer), *Athyrium Filix-foemina,* var. *clarissima, Scolopendrium vulgare,* var. *crispum-Drummondae,* and *Marsilia* (Strasburger); also in some Phanerogams, viz. Compositae (*Taraxacum,* Murbeck; *Antennaria alpina,* Juel; sp. of *Hieracium* (Rosenberg): Rosaceae: Ranunculaceae (*Thalictrum purpurascens,* Overton).

Many other ferns are known to be apogamous, but they are not included here because the details of their nuclear structure have not been investigated. APosPoRY. - The known modes of apospory may be arranged as follows: *Pseudapospory:*

a spore is formed but without meiosis, so that it is diploid - observed only in heterosporous plants, viz. certain species of *Marsilia* (*e.g. Marsilia Drummondii*) where the megaspore has a diploid nucleus (32 chromosomes) and the resulting prothallium and female organs are also diploid (Strasburger); and in various Phanerogams, some Compositae (*Taraxacum* and *Antennaria alpina*, Juel), some Rosaceae (*Eu-Alchemilla*, Strasburger), and occasionally in *Thalictrum purpurascens* (Overton), where the megaspore (embryosac) is diploid; in some species of *Hieracium* it has been found (Rosenberg) that adventitious diploid embryo-sacs are developed in the nucellus: these plants are also apogamous.

Eu-apospory: no spore is formed - of this there are two varieties:

a. With meiosis: this occurs in some Thallophyta which form no spores; the sporophyte of the Fucaceae bears no spores, consequently meiosis takes place in the developing sexual organs; the Conjugate Green Algae also have no spores, meiosis taking place in the germinating zygospore which develops directly into the sexual plant.

b. Without meiosis: the gametophyte is developed upon the sporophyte by budding; that is, spore-reproduction is replaced by a vegetative process: for instance, in mosses it has been found possible to induce the development of protonema, the first stage of the gametophyte, from tissuecells of the sporogonium: similarly, in certain ferns (varieties of *A thyrium Filix-foemina, Scolopendrium vulgare, Lastraea pseudo-mas, Polystichum angulare,* and in the species *Pteris aquilina* and *Asplenium dimorphum*), the gametophyte (prothallium) is developed by budding on the leaf of the sporophyte, and in some of these cases it has been ascertained that the gametophyte so developed has the same number (2x) of chromosomes in its nuclei as the sporophyte that bears it - that is, it is diploid.

Apospory has been found to be frequently associated with apogamy; in fact, in the absence of meiosis, this association would appear to be inevitable.

Combined Apospory and Apogamy. - Instances have been given of the occurrence of both apospory and apogamy in the same life-history; but in all of them there is a regular succession of sporophyte and gametophyte. The cases now to be considered are those in which one or other of the generations gives rise directly to its like, sporophyte to sporophyte, gametophyte to gametophyte, the normally intervening generation being omitted.

It is possible to conceive of this abbreviation of the life-history taking place in various ways. Thus, a sporophyte might be developed from a haploid spore instead of a gametophyte as is the normal case, but this has not been observed: again, a sporophyte might be developed from a diploid spore (as distinguished from a zygote or a diploid oosphere), a possibility that is to some extent realized in the life-history of some Uredineae in which successive forms of the polymorphic sporophyte are developed from diplogonidia. Similarly a gametophyte might be developed from a fertilized or an unfertilized female cell: the latter possibility is to some extent realized in those Algae (e.g. *Ulothrix, Ectocarpus*) in which the sexual cells (isogametes), if they fail to conjugate, germinate independently as gonidia, giving rise to gametophytes.

The more familiar mode is that of vegetative budding, as already mentioned. When a "viviparous" fern or Phanerogam reproduces itself by a bud or a bulbil, both spore-formation and the sexual act are passed over: sporophyte springs from sporophyte. Remarkable cases of this have been observed in certain Phanerogams (*Coelebogyne ilicifolia, Funkia ovata, Nothoscordum fragrans, Citrus,* sp. of *Euonymus, Opuntia vulgaris*) in the ovule of which adventitious embryos are formed by budding from cells of the nucellus: with the exception of *Coelebogyne,* it appears that this only takes place after the oosphere has been fertilized. In other plants it is the gametophyte that reproduces itself by means of gemmae or bulbils, as commonly in the Bryophyta, the prothallia of ferns, &c.

The abnormalities described are all traceable to reproductive degeneration; the final result of which is that true reproduction is replaced more or less completely by vegetative propagation. It may be inquired whether degeneration may have proceeded so far in any plant of sufficiently high organization to present spore-formation, or sexual reproduction, or both, as to cause the plant to reproduce itself entirely and exclusively by the vegetative method. The only such case that suggests itself is that of *Caulerpa* and possibly some other Siphonaceous Green Algae. In this plant no special reproductive organs have yet been discovered, and it certainly reproduces itself by the breaking off of portions of the body which become complete plants: but it is quite possible that reproductive organs may yet be discovered.

Physiology of Reproduction

The reproductive capacity of plants, as of animals, depends upon the fact that the whole or part of the protoplasm of the individual can develop into one or more new organisms in one or other of several possible ways. Thus, in the case of unicellular plants, the whole of the protoplasm of the parent gives rise, whether by simple division or otherwise, to one or more new plants. Reproduction necessarily closes the life of the individual: here, as August Weismann long ago pointed out, there is no natural death, for the whole of the protoplasm of the parent continues to live in the progeny. In multicellular plants, on the contrary, the reproductive function is mainly discharged by certain parts of the body, the reproductive organs, the remainder of the body being essentially vegetative - that is, concerned with the maintenance of the individual. In these plants it is only a part of the protoplasm that continues to live in their progeny; the remainder, the vegetative part, eventually dies.

It is therefore possible to distinguish in them, on the one hand, the essentially reproductive protoplasm, which may be designated by Weismann's term *germ-plasm,* though without necessarily adopting all that his use of it implies, and the essentially vegetative, mortal protoplasm, the *somatoplasm,* on the other. In the unicellular plant no such distinction can be drawn, for the whole of the protoplasm is concerned in reproduction. But even in the most highly organized multicellular plant this distinction is not absolute: for, as already explained, plants can, in general, be propagated by the isolation of almost any part of the body, that is vegetatively, and this implies the presence of germ-plasm elsewhere than in the special reproductive organs.

If the attempt be made to distinguish between the organs of vegetative propagation and those of true reproduction, the nearest approach would be the statement that the former contain both germ-plasm and somatoplasm, whereas the latter, or at least the reproductive cells, consist entirely of germ-plasm.

The question now arises as to the exact seat of the germ-plasm, and the answer is to be looked for in the results of the numerous researches into the structure and development of the reproductive cells that form so large a part of the biological work of recent years. The various facts already mentioned suffice to prove that the nucleus plays the leading part in the reproductive processes of whatever kind: the general conclusion is justified that no reproductive cell can develop into a new organism if deprived of its nucleus. It may be inferred that the nucleus either actually contains the germ-plasm, or that it controls and directs the activities of the germ-plasm present in the cell. It is not improbable that both these inferences may be true. At any rate there is no sufficient ground for excluding the cooperation of the cytoplasm, especially of that part of it distinguished as *kinoplasm,* in the reproductive processes.

Pursuing the ascertained facts with regard to the nucleus, it is established that the part of it especially concerned is the linin-network which consists of the chromosomes. The behaviour, as already described, of the chromosomes in the various reproductive processes has led to the conclusion that the hereditary characters of the parent or parents are transmitted in and by them to the progeny: that they constitute, in fact, the material basis of heredity. They can hardly, however, be regarded as the ultimate structural units, for the simple reason that their number is far too small in relation to the transmissible characters. It has been suggested (Farmer) that the chromomeres are the units, but the number of these would seem to be hardly sufficient. It seems necessary to fall back upon hypothetical ultimate particles, as suggested by Darwin, de Vries and Weismann, which may be generally termed *pangens*. The chromomeres may be regarded as aggregates of such particles, the "ids" of Weismann.

The foregoing considerations make it possible to attempt an explanation of the various reproductive processes.

VEGETATIVE PROPAGATION

It is easily intelligible that the two individuals produced by the division of a unicellular plant should resemble the parent and each other; for, the division of the parent-nucleus being homotypic, the chromosomes which go to constitute the nucleus of each daughter-cell are alike both in number and in nature, and exactly repeat the constitution of the parent-nucleus.

In the more complicated cases of propagation by bulbils, cuttings, &c., the development of the new individual, or of the missing parts of the individual (roots, &c.), may be ascribed to the presence in the bulbil or cutting of the necessary pangens.

REPRODUCTION BY GONIDIA

In this case a single cell gives rise to a complete new organism resembling the parent. The inference is that the gonidium is a portion of the parental germplasm, in which all the necessary pangens have been accumulated.

REPRODUCTION BY SPORES

In this case, also, an entire organism is developed from a single cell, but with this peculiarity that the resulting organism is unlike that which bore the spore, a peculiarity which has not yet been explained. It has been already stated that the development of true spores involves meiosis, and this process is no doubt related to the behaviour of the spore on germination; but the nature of this relation remains obscure. It might be assumed that, as the result of meiosis, the nucleus of the spore receives only gametophytic pangens. But the assumption is rendered impossible by the fact that the spore gives rise to a sexual organism, the reproductive cells of which, after the sexual act, produce a sporophyte. Clearly sporophytic pangens must be present as well in the spore as in the gametophyte and in its sexual cells. It can only be surmised that they exist there in a latent condition, dominated, as it were, by the gametophytic pangens.

SEXUAL REPRODUCTION

Here, again, as yet unanswered questions present themselves. The essence of a sexual cell is that it cannot give rise by itself to a new organism, it is only truly reproductive after the sexual act: this peculiarity is just what constitutes its sexuality. Minute investigation has not yet detected any essential structural difference between a sexual cell and a spore; on the contrary, the results so far obtained have established that they essentially agree in being post-meiotic (haploid). Why then do they differ so fundamentally in their reproductive capacities? Again, sexual cells differ in sex; but there are as yet no facts to demonstrate any essential structural difference between male and female cells. What is known about them tends to prove their structural similarity rather than their difference. But it is possible that their difference may be chemical, and so not to be detected by the microscope.

The normal *sexual act* has been described as consisting in the fusion, first, of two cells, then of their nuclei, and finally, often after a long interval, of their chromosomes and of their chromomeres in meiosis. What causes determined these fusions is a question that is only partly answered. It is known in certain cases (e.g. ferns and mosses) that the male cell is attracted to the female by chemical substances secreted for the purpose by the female organ; that it is a case of *chemio-.taxis.* Probably this is more common than experiment has yet shown it to be. It is quite conceivable that the consequent cell-fusion, as also the subsequent fusions of nuclei and of chromosomes, are likewise cases of chemiotaxis, depending upon chemical differences between the fusing structures.

The sexual process çan only take place between cells which are related to each other in a certain degree; that is, it depends upon *sexual affinity*. It is the general rule that it takes place between cells derived from different individuals of the same species; that is, *cross-fertilization* is the rule. This is necessarily the case when the male and female organs are developed upon different individuals, when the plant is said to be *dioecious.* When both kinds of organs are developed upon the same individual (*monoecious*), self-fertilization may and often does occur; but it is commonly hindered by various special arrangements, of which *dichogamy* is the most common; that is, that the male and female organs are not mature at the same time. But though these arrangements favour cross-fertilization, they do not absolutely prevent selffertilization. In some cases, *cleistogamic* flowers, for instance, self-fertilization alone is possible. The general conclusion is that though cross-fertilization is the more advantageous form of sexual reproduction, still self-fertilization is more advantageous to the species than no fertilization at all.

In considering this subject, it must be borne in mind that the terms used have different meanings when applied to certain heterosporous plants from those which they convey when applied to isosporus plants. In the latter cases their meaning is direct and simple: in the former it is indirect and somewhat complicated. In heterosporous plants generally the actual sexual organs are never borne upon the same individual, there is always necessarily a male and a female gametophyte; so that, strictly speaking, self-fertilization is impossible. But in the Phanerogams, where there is a process preliminary to fertilization, that of *pollination,* which is unknown in other plants, the terms and the conceptions expressed by them are applied, not to the real sexual organs, but to the spores. Thus a dioecious Phanerogam is one in which the microspores are developed by one individual, the megaspores by another; and again, self-fertilization is said to occur when the microspores (pollen) fall upon the stigma of the same flower; but this is really only *self-pollination*. To return to the sexual process itself. Whatever its nature, two sets of results follow upon the sexual act - (r) a zygote is formed, which is capable of developing into a new organism, from two cells, neither of which could so develop; (2) the hereditary sporophytic characters of the two parents are possessed by the organism so developed. These two results will now be considered in some detail.

1. *The Relation between the Sexual Act and Reproductive Capacity*: *In* the early days of the discovery of the sexual process, it was thought that the capacity for development imparted to the female cell was to be attributed to the doubling of its nuclear substance by the fusion with the male cell. Reproductive capacity does not, however, depend upon the bulk of the nuclear substance, for a spore, like an unfertilized female cell, contains but the x number of chromosomes, and yet it can give rise to a new organism. Again, it has been observed (Winkler) that a non-nucleated fragment of an oosphere of *Cystoseira* (Fucaceae) can be "fertilized" by a spermatozoid and will then grow and divide to form a small embryo, though it necessarily contains only the x number of chromosomes. From this it would appear that some stimulating influence had been exerted by the male

cell, and it is probably in this direction that the desired explanation is to be sought. Some important confirmatory facts have been recorded with regard to certain animals (sea-urchins). It has been observed (Loeb) that treatment with magnesium chloride will cause the ova to grow and segment; and similar results have been obtained (Winkler) by treating the ova with a watery extract of the male cells. Hence it may be inferred that the male cell carries with it, either in its cytoplasm (kinoplasm), or in its nucleus, extractable substances, perhaps of the nature of enzymes, that stimulate the female cell to growth.

It may be mentioned that the stimulating effect of fertilization is not necessarily confined to the female cell; very frequently adjacent tissues are stimulated to growth and structural change. In a Phanerogam, for instance, the whole ovule grows and develops into the seed: the development of endosperm in the embryo-sac is initiated by another nuclear fusion, taking place between the second male nucleus and the endosperm-nucleus: the ovary, too, grows to form the fruit, which may be dry and hard or more or less succulent: the stimulating effect may extend to other parts of the flower; to the perianth, as in the mulberry; to the receptacle, as in the strawberry and the apple: or even beyond the flower to the axis of the inflorescence, as in the fig and the pine-apple. Analogous developments in other groups are the calyptra of the Bryophyta, the cystocarps of the Red Algae, the ascocarps of the Ascomycetes, the aecidia of the Uredineae, &c.

2. *The Relation of the Sexual Act to Heredity*: *The* product of the sexual act is essentially a diploid cell, the zygote, which actually is or gives rise to a sporophyte. The sexual heredity of plants consequently presents the peculiar feature that the organism resulting'; from the sexual act is quite

> unlike its immediate parents, which are both gametophytes. But it is clear that the sporophytic characters must have persisted, though in a latent condition, through the gametophyte, to manifest themselves in the organism developed from the zygote.

The real question at issue is as to the exact means by which these characters are transmitted and combined in the sexual act. There is a considerable amount of evidence that the hereditary characters are associated with the chromomeres, and that it is rather their limn-constituent than their chromatin which is functional (Strasburger): that they constitute, in fact, the material basis of heredity. From this point of view it is probable that the last phase of the sexual act, the fusion of the chromomeres in meiosis, represents the combination of the two sets of parental characters. What exactly happens in the pseudo-chromosome stage is not known; at any rate this stage offers an opportunity for a complete redistribution of the substance of the chromomeres - in other words, of the parental pangens. It is a striking fact that, in the subsequent nuclear division, the distribution of the chromosomes derived from the male and female parents (when they can be distinguished) seems to be a matter of indifference: they are not equally distributed to the two daughter-nuclei. The explanation would appear to be this, that they are not any longer male and female as they were before meiotic fusion; and that it is because they now contain both male and female nuclear substance that their equal distribution to the daughter-nuclei is unimportant.

The nature of this redistribution of the substance of the chromomeres is still tinder discussion. Some regard it as essentially a chemical process, resulting in the formation of new compounds: others consider it to be rather a physical process, a new material system being formed in the rearrangement of the pangens; here it must be left for the present.

The various ways in which the parental characters manifest themselves in the progeny are fully dealt with in the articles Heredity, Hybridism, Mendelism. It will suffice to say that the progeny, though maintaining generally the characters of the species, do not necessarily exactly resemble either of the parents, nor do they necessarily present exactly intermediate characters: they may vary more or less from the type. It is an interesting fact, the full significance of which has not yet been worked out, that, as a rule, plants that vary profusely are those in which the characteristic 2X number of chromosomes is high.

Brief reference may be made to the cases of abnormal sexual or pseudo-sexual reproduction described above under *Apogamy*. Taking first the cases of true apogamy, there is clearly no need for any sexual process, for, since no meiotic division has taken place, the gametophyte is diploid; its cells, whether vegetative or contained in female organs, possess the capacity for both development and the transmission of the sporophytic characters. It is not remarkable that such a gametophyte should be able to give rise directly to a sporophyte; but it is remarkable, in the converse case of apospory, that a sporophyte should give rise to a diploid gametophyte rather than to another sporophyte. In the latter case the tendency to the regular development of the alternate form appears to override the influence of the diploid nucleus.

Turning to the various forms of pseudo-apogamy, there are first those in which fusion takes place between two apparently female organs (some Uredineae; Christman), and those in which it takes place between nuclei within the same female organ (*Humaria;* Blackman). If these are to be regarded physiologically as sexual acts, it must be inferred that the fusing organs or nuclei have come to differ from each other to some extent; for it is unthinkable that equivalent female organs or cells should be able to fertilize, or to be fertilized by, one another. There are finally those cases in which apparently vegetative cells take part in the sexual act, as in *Phragmidium*

(Blackman), where the female organ fuses with an adjacent vegetative cell, and in the fern-prothallium (Farmer), where the nuclei of two vegetative cells fuse. They would seem to indicate that vegetative cells may, in certain circumstances, contain sufficient germ-plasm to act as sexual organs without being differentiated as such.

An interesting question is that of the origin of apogamy. It is no doubt the outcome of sexual degeneration; but this general statement requires some explanation. In certain cases apogamy seems to be the result of the degeneration of the male organ; as in *Humaria,* where there is no male organ, and in *Lachnea,* where the male organ is rudimentary. In others, as in the Uredineae, it is apparently the female organ that has degenerated, losing its receptive part, the trichogyne; the male cells (spermatia) are developed normally, and there is no reason to believe that they might not fertilize the female organ were there the means of penetrating it. In yet other cases the degeneration occurs at a different stage in the life-history, in the development of the spores. In the apogamous ferns investigated, meiosis is suppressed and apogamy results. In the heterosporous plants which have been investigated (e.g. *Marsilia, Eu-Alchemilla*) it has been observed that the microspores are so imperfectly developed as to be incapable of germinating, so that fertilization is impossible; and it is perhaps to this that the occurrence of apogamy is to be attributed. This abnormal development of the spores may be regarded as a variation; and in most cases it occurs in plants that are highly variable and often have a high 2X number of chromosomes.

It will be observed that such physiological explanation as can be given of the phenomena of reproduction is based upon the results of the minute investigation of the changes in nuclear structure associated with them. The explanation is often rather suggested than proved, and some fundamental facts still remain altogether unexplained. But it may be anticipated that a method of research which has already so successfully justified itself will not fail in the future to elucidate what still remains obscure.

AGRICULTURE

Techniques aimed at crop improvement have been utilized for centuries. Today, applied plant science has three overall goals: increased crop yield, improved crop quality, and reduced production costs. Biotechnology is proving its value in meeting these goals. Progress has, however, been slower than with medical and other areas of research. Because plants are genetically and physiologically more complex than single-cell organisms such as bacteria and yeasts, the necessary technologies are developing more slowly.

IMPROVEMENTS IN CROP YIELD AND QUALITY

In one active area of plant research, scientists are exploring ways to use genetic modification to confer desirable characteristics on food crops. Similarly, agronomists are looking for ways to harden plants against adverse environmental conditions such as soil salinity, drought, alkaline earth metals, and anaerobic (lacking air) soil conditions.

Genetic engineering methods to improve fruit and vegetable crop characteristics - such as taste, texture, size, colour, acidity or sweetness, and ripening process, are being explored as a potentially superior strategy to the traditional method of cross-breeding.

Research in this area of agricultural biotechnology is complicated by the fact that many of a crop's traits are encoded not by one gene but by many genes working together. Therefore, one must first identify all of the genes that function as a set to express a particular property. This knowledge can then be applied to altering the germlines of commercially important food crops. It will be possible to transfer the genes regulating nutrient content from one variety of tomatoes into a variety that naturally grows to a larger size. Similarly, by modifying the genes that control ripening, agronomists can provide supplies of seasonal fruits and vegetables for extended periods of time.

Biotechnological methods for improving field crops, such as wheat, corn and soybeans, are also being sought, since seeds serve both as a source of nutrition for people and animals and as the material for producing the next plant generation. By increasing the quality and quantity of protein or varying the types in these crops, we can improve their nutritional value. A major protein of corn has very little of two amino acids, lysine and tryptophan, which are essential for human growth. Increased amounts of these amino acids could make corn products a source of improved protein.

BIOPESTICIDES AND BIOFERTILIZERS

Biotechnology makes it possible to develop bacteria essential to herbicide and other pesticide compounds. Certain chemicals produced by these organisms are called allelopathic agents. These chemicals act as natural herbicides, preventing the growth of other plant species in the same geographic area. Black walnut trees, release an allelopathic agent against tomato plants.

Modern high-yield agriculture entails consumption of vast amounts of chemicals for use as fertilizers and as agents to control pests and plant diseases, and any means that will permit the plant to do this work itself could result in significant savings for the farmer. Soybeans and certain other legumes produce their own source of usable nitrogen fertilizer by a process known as nitrogen fixation. This process is made possible by a bacterium that grows symbiotically on the plant's roots. In the nitrogen-fixing process, microbes capture atmospheric nitrogen and biochemically convert it into water-soluble nitrogen. This form of nitrogen is an essential nutrient for increasing the quantity and quality of plant yield.

This bacteria will not assist in the growth of other important crops, such as corn and cereal plants. But research on nitrogen-fixing bacteria and legumes may show how we can modify either the bacteria or non-leguminous plants, thereby making many crops more nearly self-sufficient in obtaining nitrogen.

Biotechnology is central to the search for effective, environmentally safe and economically sound alternatives to chemical pesticides. Biotechnology may be used to protect commercial crop plants from insect pests and promises to guard against further environmental deterioration and to provide a useful alternative to traditional methods of insect pest control.

Biopesticides degrade rapidly in the environment - a major environmental benefit. The active elements of bacterial pesticides are proteins that are fragile molecules. Once exposed to the sun and other natural elements, these proteins are quickly broken down, thereby prohibiting spread to groundwater and other animal and plant species. This will help keep our water supply safe to drink, our lakes and streams habitable for water life and recreation.

Unlike chemical pesticide technology, biopesticide technology is based on potent, naturally occurring proteins. These living particles are produced in nature by microorganisms such as Bacillus thuringiensis (B.t.). Discovered at the turn of the century, B.t. has been used without risk in the United States for almost three decades by home gardeners, farmers, and forestry officials. Its active component, a protein, specifically attacks the stomachs of target pests, disrupting their digestive tracks so thoroughly that the pests stop eating and eventually die of starvation. Higher organisms, such as mammals, fish, birds, and other non-target species remain unthreatened, however, because their stomach acid easily breaks down the protein toxin.

The delivery of these biopesticides varies in method and design. In one method, dormant spores of B.t. are dusted on crops. The spores then become active and multiply, covering plants with a bacteria poisonous to the target insects that feed on them. The B.t. toxin gene can also be inserted into the genetic makeup of crops, giving them a built-in resistance to insects. Similarly, the toxin gene can be put into a third party, such as a microorganism that lives within the plant's sap. These organisms - known as endophytes - multiply within the host plant and move throughout the plant's vascular system, forming a microscopic defence against feeding insects. This process resembles vaccines moving throughout a person's vascular system to defend against harmful disease.

Some of the concerns farmers raise about having to use increasingly dangerous pesticides to produce adequate crops may well be addressed by biotechnology. Further research in agricultural biotechnology and biopesticide development aims to provide attractive alternatives to the farmer that will lower overall unit cost of production and allow the farmer to be more competitive in the highly cost-sensitive world markets. While some uses of chemical pesticides will be necessary for decades to come, continued development by biotechnology companies of useful biological pesticides will offer farmers viable alternatives.

CLONING OF PLANT CELLS AND MANIPULATION OF PLANT GENES

Plant cells exhibit a variety of characteristics that distinguish them from animal cells. These characteristics include the presence of a large central vacuole and a cell wall, and the absence of entioles, which play a role in mitosis, meiosis, and cell division. Along with these physical differences, another factor distinguishes plant cells from animal cells, which is of great significance to the scientist interested in biotechnology: Many varieties of full-grown adult plants can regenerate from single, modified plant cells called protoplasts - plant cells whose cell walls have been removed by enzymatic digestion. More specifically, when some species of plant cells are subjected to the removal of the cell wall by enzymatic treatment, they respond by synthesizing a new cell wall and eventually undergoing a series of cell divisions and developmental processes that result in the formation of a new adult plant. That adult plant can be said to have been cloned from a single cell of a parent plant.

Plants that can be cloned with relative ease include carrots, tomatoes, potatoes, petunias, and cabbage, to name only a few. The capability to grow a whole plant from a single cell means that researchers can engage in the genetic manipulation of the cell, let the cell develop into a completely mature plant, and examine the whole spectrum of physical and growth effects of the genetic manipulation within a relatively short period of time. Such a process is far more straightforward than the parallel process in animal cells, which cannot be cloned into full-grown adults. Therefore, the results of any genetic manipulation are usually easier to examine in plants than in animals.

A CLONING VECTOR THAT WORKS WITH PLANT CELLS

Not all aspects of the genetic manipulation of plant cells are readily accomplished. Not only do plants usually have a great deal of chromosomal material and grow relatively slowly as compared with single cells grown in the laboratory, but few cloning vectors can successfully function in plant cells. While researchers working with animal cells can choose among a wide variety of cloning vectors to find just the right one, plant cell researchers are currently limited to just a few basic types of vectors.

Perhaps the most commonly used plant cloning vector is the "Ti" plasmid, or tumor-inducing plasmid. This plasmid is found in cells of the bacterium known as Agrobacterium tumefaciens, which normally lives in soil. The bacterium has the ability to infect plants and cause a crown gall, or tumorous lump, to form at the site of infection. The tumor-inducing capacity of this bacterium results from the presence of the Ti plasmid. The Ti plasmid itself, a large, circular, double-stranded DNA molecule, can replicate independently of the A. tumefaciens genome. When these bacteria infect a plant cell, a 30,000 base-pair segment of the Ti plasmid - called T DNA - separates from the plasmid and incorporates into the host cell genome. This aspect of Ti plasmid function has made it useful as a plant cloning vector.

The Ti plasmid can be used to shuttle exogenous genes into host plant cells. This type of gene transfer requires two steps: 1) the endogenous, tumor-causing genes of the T DNA must be inactivated and, 2) foreign genes must be inserted into the same region of the Ti plasmid. The resulting recombinant plasmid, carrying up to approximately 40,000 base pairs of inserted DNA and including the appropriate plant regulatory sequences, can then be placed back into the A. tumefaciens cell. That cell can be introduced into plant cell protoplasts either by the process of infection or by direct insertion.

Once in the protoplast, the foreign DNA, consisting of both T DNA and the inserted gene, incorporates into the host plant genome. The engineered protoplast - containing the recombinant T DNA - regenerates into a whole plant, each cell of which contains the inserted gene. Once a plant incorporates the T DNA with its inserted gene, it passes it on to future generations of the plant with a normal pattern of Mendelian inheritance.

One of the earliest experiments that involved the transport of a foreign gene by the Ti plasmid involved the insertion of a gene isolated from a bean plant into a host tobacco plant. Although this experiment served no commercially useful purpose, it successfully established the ability of the Ti plasmid to carry genes into plant host cells, where they could be incorporated and expressed.

A. TUMEFACIENS INFECTS A LIMITED VARIETY OF PLANT TYPES

The fact that only certain types of plants were naturally susceptible to infection with the host bacterial organism initially limited the usefulness of the Ti plasmid as a cloning vector. In nature, A. tumefaciens infects only dicotyledons or "dicots" - plants with two embryonic leaves. Dicotyledenous plants, divided into approximately 170,000 different species, include such plants as roses, apples, soybeans, potatoes, pears, and tobacco. Unfortunately, many important crop plants, including corn, rice, and wheat, are monocotyledons - plants with only one embryonic leaf - and thus could not be easily transfected using this bacterium.

OVERCOMING THE LIMITED RANGE OF A. TUMEFACIENS INFECTION

Research efforts in the past few years have reduced the limitations of A. tumefaciens. Scientists discovered that by using the processes of microinjection, electroporation, and particle bombardment, naked DNA molecules can be introduced into plant cell types that are not susceptible to A. tumefaciens transfection.

Microinjection involves the direct injection of material into a host cell using a finely drawn micropipette needle. Electroporation uses brief pulses of high voltage electricity to induce the formation of transient pores in the membrane of the host cell. Such pores appear to act as passageways through which the naked DNA can enter the host cell. Particle bombardment actually shoots DNA-coated microscopic pellets through a plant cell wall. These developments, important in the commercial application of plant genetic engineering, render the valuable food crops of corn, rice, and wheat susceptible to a variety of manipulations by the techniques of recombinant DNA and biotechnology.

GENETIC VARIATION

Our main interest is to isolate genes that are responsible for the intra- and interspecific morphological variation in plants. Therefore, the analysis of natural genetic variation is the starting point for the majority of individual projects in our lab.

For the isolation of genes involved in morphological variation, exploiting natural genetic variation offers some distinct advantages as compared to classical mutagenesis experiments. Most importantly, it allows the identification of naturally occurring, differentially active gene alleles that are likely targets for the evolution of morphological variation. Because naturally occurring lines do not carry major deleterious mutations that impair their survival in the wild, the approach also counter-selects against the isolation of mutant alleles that result in strong phenotypes as commonly detected in mutagenesis experiments. This applies in particular to wild populations with a high level of inbreeding, such as in our favourite model organism, *Arabidopsis thaliana*.

So why has this approach not been followed earlier? Mainly because the isolation of genes that modify quantitative traits has technically not been feasible. However, the rich biological and genomic resources, such as the fully sequenced genome, have turned Arabidopsis into an ideal model system to investigate the molecular basis of natural genetic variation at the gene level.

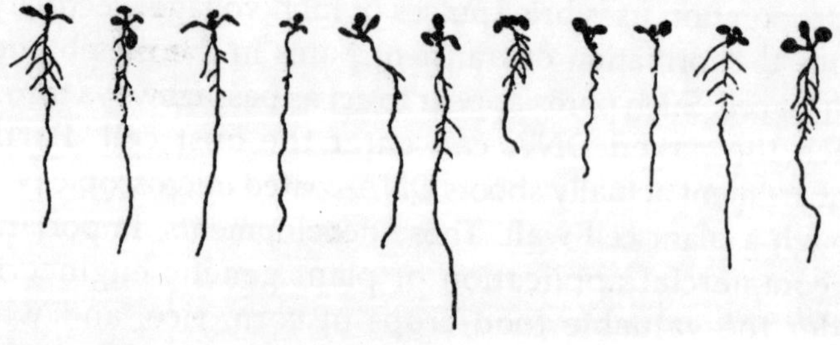

Natural morphological variation of root system architecture between wild isolates of Arabidopsis. Most of this variation results from the influence of genetic factors.

Plant organs are formed in a continuous, post-embryonic manner by ordered cell divisions and expansions. The extent of growth ultimately determines organ shape. Thus, we are primarily interested in genes that modulate growth rate. To isolate such genes, we focus on the root system. This is because of our interest in root biology per se, but also because roots grow from a so-called meristematic region at their tip, which produces concentric tissue layers through controlled cell proliferation, elongation and differentiation. This feature enables us to reduce the problem of accurately measuring growth from three dimensions to a single one. Nevertheless, analysis of root system growth requires care and highly reproducible growth conditions, because root system development is very plastic and responds to many macro- and micro-environmental cues. The natural variation in root system morphology between Arabidopsis accessions collected from different locations and grown in identical growth chamber conditions on tissue culture media.

So far, we have isolated an important regulator of root system architecture from the accession Uk-1. This gene is a major so-called quantitative trait locus (QTL), which we gave the name *Brevis Radix (BRX),* latin for "short root". Other efforts in this area include attempts to isolate two other QTLs for root growth located in other accessions.

PLANT HORMONE ACTION

Plant hormones are at the heart of many plant developmental processes. They determine critical growth parameters, such as the rate of cell division or elongation, as well as developmental decisions, for instance where along a primary root a lateral root should be formed.

Different plant hormones control different developmental processes to varying degrees, sometimes in the same, sometimes in opposite directions. Whether and how the overlapping influence of different hormones on traits is coordinated is largely unknown. Part of the research in our lab aims to determine how the major plant hormone auxin, arguably the most important in shaping the plant, regulates downstream events that lead to discrete morphological change. After all, those changes require structural rearrangements on the cellular and organ level after hormone signaling. One tool to answer this question are the *hy5* and *hyh* mutants of Arabidopsis, which display constitutively increased auxin signaling.

Another important question is the nature of crosstalk between hormone pathways, that is the impact of more than one hormone signaling pathway on the same trait or molecular event.

Much to our surprise, we isolated an important regulator of hormone pathway interaction in the root through our natural variation project: loss of function of the *BRX* gene results in reduced root growth because *BRX* is required for the biosynthesis of the hormone brassinolide, whose concentration is in turn rate-limiting for auxin action. Since *BRX* is itself under auxin control, this creates a feedback loop required for optimal root growth. Whether an equivalent loop works in the shoot, and whether this feedback is conserved in other species are some of the questions we are currently interested in with regard to this project, generously funded by the Swiss National Science Foundation.

GENETIC VARIATIONS

Variation

If you have grown Fast Plants you will have noticed variation among the individual plants within the group. Variation can range from a little to a lot. This article is designed to help teachers and students understand the basis of the variation they observe in their plantings.

Variation is one of the fundamental characteristics of life. All organisms exhibit some variation among individuals. Understanding the ways that variation is manifested in organisms, how it comes to be expressed through the development of the individual, and how it is transmitted from one individual to the next generation of individuals are central themes in biology.

Working with Fast Plants will enrich a student's understanding of variation. By observing the growth and development of a Fast Plants through the various stages in the life cycle, students will become aware of many visible features, or phenotypes, that make up the organism.

Only, however, upon close observation of a population of plants, will they become aware that the characteristics observed on one plant vary more or less on other plants. Such is the nature of variation.

Phenotypic Variation

Phenotypic variation, e.g. plant height at a particular stage of development, is considered to be the expression of the genetic makeup (genotype) of the individual as it interacts with the environment. Variation in plant height among individuals in a population is therefore due to variation in the interaction between the genotype and the environment, as expressed through the development of each individual plant.

The interrelationships suggested should provide students with numerous ideas for experimentation to probe the basis for the variation observed among their individual Fast Plants.

Describing and Observing Phenotypic Variation

In order to be useful in an experiment the phenotype must be described using terms that are widely understood and easily communicated. For these reasons scientists have agreed upon various standards or descriptors to describe characteristics in the natural world. Descriptors take many forms. The choice of how to describe what you observe is important, because it will determine the kinds of descriptors used and establish the basis for recording, analyzing and communicating results.

Environmental Variation

Much can be learned about the role of light, temperature, and nutrients on plant development from experiments in which one or more environmental parameters are changed.

Because Fast Plants are highly responsive to changes in the environment, they are ideal for examining the role of the environment on the expression of phenotypic variation.

Although Fast Plants are able to grow within a wide range of environmental conditions, for most investigations it is recommended that they be grown under uniform and ideal conditions. In this way variation arising from sub-optimal conditions of environment will be minimized. The Wisconsin Fast Plants Information Document (WFPID) Understanding the Environment describes how to provide and maintain the various physical, chemical and biotic components of the environment that are most suitable for Fast Plants.

Genotypic Variation

How can students use Fast Plants to investigate the contribution of the genotype to the phenotype? Fast Plants, rapid cycling Brassica rapa, are genotypically variable in that they have a genetically controlled mating system that prevents self-fertilization and favours out-crossing among individuals. As a consequence, even seed stocks selected for uniformity of specific phenotypes and genotypes exhibit considerable variation for other traits.

The Wisconsin Fast Plants Programme has developed a number of genetic stocks of rapid cycling Brassica rapa for genetic investigations on the nature and inheritance of variation. Some WFP stocks contain distinctive mutant phenotypes, e.g., anthocyaninless plant, anl, yellow green plant, ygr1, rosette, ros, and male sterile, mst2, that are suitable for Mendelian genetics.

Other stocks exhibit phenotypes whose expression may vary continually and which may be quantified as discrete or countable units, e.g., number of hairs, or as estimates of size, e.g. petite dwarf, dwf1, or of intensity of colour saturation, e.g., purple anthocyanin. These quantitative phenotypes may be conditioned by a few or many genes and normally require numerical descriptions in which the statistical notations of population size, (n), range (r), arithmetic mean (x), and standard deviation (s) are applied.

Yet other stocks have been developed to combine both simply inherited mutant genotypes and quantitatively expressed phenotypes. An important part of WFP is the continuing development and improvement of seed stocks for uses in genetics.

PEDIGREE

Pedigree-based whole genome (PBWG) mapping offers significant potential to increase the genetic gain made in plant breeding programmes through the development of systems for marker assisted selection. In in-breeding plants, the identification of genomic regions associated with traits of importance traditionally uses populations derived from single crosses of inbred lines. This strategy however is frequently sub-optimal for direct application in breeding programmes for numerous reasons. The QTL (quantitative trait loci) - trait association strategies frequently use obscure or out-dated parents. This strategy competes directly with breeding programmes for significant resources for phenotypic evaluation. It is also limited in coverage because it addresses only the genes segregating in the populations under study. Additionally this strategy provides no information about the frequency or value of the QTLs in the breeding populations and hence requires a further step of validating the usefulness of markers for direct application in the breeding programme. Due to the wide-spread adoption of this QTL analysis approach with specific doubled haploid populations, marker application is being impeded through the lack of information about the polymorphism and marker-trait association in genetic material relevant to Australian northern region cereal improvement.

Pedigree-based whole genome marker application provides a vehicle for incorporating marker technologies into applied breeding programmes by bridging the gap between development and implementation. The PBWG marker concept uses pedigree information to identify markers linked to traits based on identity by descent in breeding populations. The approach makes efficient use of pedigree, phenotypic and genotypic information collected in the normal breeding programme, thereby exploiting this valuable resource. The PBWG approach integrates very closely with marker assisted recurrent parent recovery (MARPR), a technology already applied to many crops including barley, maize and rice, that can speed variety development in back-crossing strategies.

Effective use of a PBWG-MARPR marker approach requires relevant pedigree and phenotypic data, low-cost, high throughput genotyping and a data management and analysis system that combines pedigree information, genotypic and phenotypic data. Implementation of these components allows linkage to advanced molecular marker and related research, which directly addresses constraints and opportunities arising within the context of the breeding programme. The current project is piloting the design and development of a system that integrates these components of PBWG and MARPR marker application for use in applied breeding programmes.

Methods

This project aims to develop the protocols and systems necessary to implement the PBWG approach using the DPI&F and EGA northern region focussed wheat and barley breeding programmes as case studies. Within each case study, the overall strategy is as follows; to identify appropriate interrelated sets of germplasm, to identify a suite of robust and polymorphic molecular markers, to fingerprint the germplasm and verify the pedigrees, to collate phenotypic, genotypic and pedigree information, to verify and repeat phenotypic information as necessary, to undertake linkage analysis, to validate the markers alternative branches of the pedigree and finally to validate the markers in un-related germplasm.

Wheat Case-Study

A core set of 126 wheat varieties and breeding lines from pedigrees relevant to the northern region with a focus on Condor and Cook related material has been catalogued. 118 SSR markers were selected for assessment in wheat spread at approximately 10cM intervals on chromosomes 2B, 2D, 3B, 3D, 4A, 4B, 4D and 7A. These chromosomes were chosen on the basis of location of genes and putative QTLs associated with height, milling yield, flour colour, black point resistance and rust resistances. In addition to the SSR markers, 63 DArT markers were applied to 93 genotypes by Triticarte P/L.

Barley Case-Study

A set of 66 barley genotypes was selected with lineage from Triumph and/or Koru), together with the other parents of the chosen advanced breeding lines. Both Triumph and Koru are European malting barleys which have contributed significantly to the development of malting and high yielding feed barleys with adaptation to the northern region of Australia. The result is a set of inter-related genotypes which are representative of a significant proportion of the Northern Barley Improvement Programme (NBIP) gene pool. Fifty-five SSR markers linked to traits of importance to the barley industry have been applied to DNA fingerprinting the 66 barley genotypes. In addition to the SSR markers, 297 polymorphic DArT markers were applied to the same genotypes by Tritcarte P/L.

Genotypic information has been obtained for 118 SSR markers from 126 wheat lines for pedigree-based genomic analysis. Additionally, information has been obtained for 63

DArT markers from 93 of the wheat lines. 297 polymorphic DArT markers have been identified and applied to the 66 barley lines, supplementing the genotypic data obtained from the 55 SSR markers. Pedigree data for the 126 wheat lines and 66 barley lines has been obtained, cleaned, validated and entered into PBMASS (Pedigree Based Marker Assisted Selection System). PBMASS is the software currently being developed for this project.

Modules currently to beta testing stage include: import, export and storage of pedigree data, printable graphical display of pedigree data including functions to calculate COI and COP, import, export and storage of molecular marker data, printable display of graphical genotypes colour coded for parental origin (using identity-by-descent) for selected combinations of genotypes and chromosomes.

Marker data will be analysed in conjunction with both existing and new phenotypic data. The wheat and barley applications will be assessed and compared to determine the effectiveness of PBWG marker analysis in producing relevant information for use in interpreting associations between traits and genomic regions under selection through the pedigrees targeted in each breeding programme and for trait selection, parent characterisation and MARPR.

The analysis of the PBGW mapping approach will allow for an enhanced understanding of the processes of selection within a breeding programme, including identification of regions under selection and will act as a basis of validation of markers and QTLs for the more effective implementation of markers, through the confirmation of QTL location, the identification of markers which will work in other, distantly related populations and in the identification of QTLs which are effective in other populations.

The expected outcomes from this project will encourage more sophisticated use of markers in targeted breeding populations as selection is applied to conserve and combine critical genomic regions for both environmental adaptation and grain and end-use quality. The use of a focussed set of genotypes means that the marker technology will be quickly available for use with a large proportion of the gene pools in the respective case study breeding programmes, however the comparison of the two case studies will also allow for the identification of critical components of this approach for implementation in other crop plants.

Chapter 3

Plant Breeding

Plant breeding, a respected discipline, has been practiced as a distinct profession for about a century. Genetic enhancement is a new term, of uncertain meaning, certainly not describing a distinct profession. What precisely is genetic enhancement? How does it differ from plant breeding? Why should we discuss it at all? The concept of genetic enhancement relates to a fairly new concept: genetic vulnerability. And genetic vulnerability is a direct by-product of successful plant breeding. Let me explain.

GENETIC ENHANCEMENT

Plant breeding as an art probably pre-dates civilization; it certainly has shown results (cultivated plant varieties) for the past 10,000 years. As a science and a full-time profession, however, plant breeding is largely a 20th century discipline. Its successes, as we all have heard many times over, have given rise to great expanses of a few highly favoured cultivars, which in turn have attracted epidemic spread of diseases and pests, uniquely adapted to their ever restricted range of hosts. From experiences with such epidemics, the concept of genetic vulnerability has been developed, along with its corollary, genetic diversity.

Promotion of genetic diversity requires introduction of new, unrelated breeding materials into basic and advanced breeding pools. Resulting new cultivars thus can display a greater diversity of genotypes, and present less opportunity, as grown on the farm, for epidemic disaster.

But to bring greater diversity into the breeding pools means going outside the range of elite adapted local materials. It requires bringing in unadapted and, thus, unproductive cultivars—or even weedy or wild species—that have only a few useful traits, usually tightly linked with a panoply of other undesirable, yield-reducing traits. Such exotic material, when crossed directly with elite adapted lines, uniformly gives unsuccessful breeding results. Useful cultivars rarely or never can be developed from such wide crosses.

But skillful breeders can overcome this problem. They do not try, immediately, to use the unadapted material for production of elite new cultivars. They settle, instead, for selection of intermediates: lines or populations that, while maintaining the good new traits from the exotics, also contain some of the germplasm of the adapted elite cultivars. The broadening alien germplasm thus is at least partly naturalized, and most importantly, the new product, genetically enhanced breeding material, now is somewhat attractive to those plant breeders who are charged primarily with the task of reliably and repeatedly turning out top quality new cultivars.

Thus, the breeder who has performed the important function of partially acclimatizing alien germplasm, while conserving its essential genetic contributions, has performed the operation of "genetic enhancement." Genetic enhancement is needed to prevent genetic vulnerability brought about by the successes of modern plant breeding.

Why is genetic enhancement (I think the plant breeders' term "prebreeding" may be more or less synonymous) thought of as a new requirement in plant breeding? Since all crop cultivars came originally from wild species, genetic enhancement—selection toward useful cultivar types—surely was needed for the original modification of the wild species. What is the difference between then and now?

The difference is that 10,000 years ago—or even 1,000 years ago—there were no obvious divisions between highly selected, elite cultivars and purposely introduced, grossly unadapted exotics. Farmer selectors worked with whatever materials they found locally, and imperceptibly, generation after generation, they selected towards the types they needed. We, of course, don't know how or if such selection was planned; we are reasonably sure, however, that the early farmer-selectors did not purposely hybridize highly select, specialized cultivars to unadapted exotics or species, with the specific aim of bringing in new traits. Thus, our farmer-selector progenitors didn't need to do "prebreeding" or genetic enhancement. From our present sophisticated point of view, that was all they did do. They continually and gradually enhanced genetic materials at hand, slowly making them more amenable to cultivation for human use.

Until sophisticated hybridization and selection schemes were devised, there were few or no sharp breaks between adapted breeding materials and newly introduced, unadapted breeding materials. Sharp breaks, if they existed, were not planned. It is true that when immigrants brought a favoured cultivar from the homeland to a new place—as to a new continent with different day-length or different climate—there was a break: the transported cultivar was unadapted to its new home until gradual selection (within the heterogeneous cultivar, or among progeny resulting from its accidental hybridization to indigenous cultivars) gave rise to new, genetically enhanced (better adapted) progeny. But such enhancement was neither planned, nor frequent.

USES OF GENETIC ENHANCEMENT

There are at least three distinct uses of genetic enhancement. The first is to prevent genetic uniformity and consequent genetic vulnerability. Only recently has pre-breeding—genetic enhancement—become a necessary, frequent, and planned part of all plant breeding activities, an essential part of germplasm diversification strategies.

Genetic enhancement has a second important purpose, that of raising yield levels to new heights. This goal is more often hoped for than achieved, but it is true that most breakthrough cultivars have highly diverse parentage. The semi-dwarf wheats, the high yield dwarf rices, the first U.S. hybrid sorghums, and even the first U.S. corn belt dent maize cultivars are examples. In each case, extensive prebreeding preceded development of the breakthrough, high-yield cultivars. The prebreeding was used to adapt diverse kinds of germplasm to new genetic backgrounds and new geographic locales.

A third use of genetic enhancement also needs mention. Genetic enhancement is used to bring in new quality traits not found in local cultivars. New levels of protein percentage in wheat or unusual starch properties in maize are examples.

WHO DOES GENETIC ENHANCEMENT?

Genetic enhancement is now necessary and useful-but it is not yet well-recognized as being so. This is evident because there is no group of scientists or professionals who call themselves "genetic enhancers", or "prebreeders"; there is no "American Society of Generic Enhancers," no "National Council of Commercial Pre-Breeders." Nor is there a recognized body of prebreeding theory. Nor can I call to mind people whose chief reputation in plant biology—basic or applied—rests on their accomplishments in prebreeding. Neither fame nor fortune—nor even generous funding—comes to those who do only prebreeding.

Yet the job gets done-exotic germplasm is incorporated into adapted, elite stocks, and new cultivars often do contain significant segments of exotic germplasm, giving new kinds of disease and insect resistance, new levels or kinds of stress tolerance, or greater yield capability.

By and large, the job of generic enhancement is done by a special subset of plant breeders and geneticists in the public sector, scientists who either know they must pre-breed in order to make further progress in variety development of their crop, or know they like the challenge of finding, incorporating, and adapting useful germplasm from unlikely places. But in most cases, the main basis of their professional reputation, and of their ability to attract research funds, has been their successful production of finished cultivars. These scientists have supported their avocation, prebreeding, with their vocation, cultivar production.

An important deviation from this rule, that prebreeders also are cultivar developers, may be arising in the international research centers such as the International Rice Research Institute (IRRI), the International Crops Research Institute for the Semi-Arid Tropics (ICRISAT) or the International Maize and Wheat Improvement Center (CIMMYT). Such centers were formed with the express purpose of making cultivars and advanced breeding materials for use in developing countries. Today, as national breeding programmes are maturing in most developing countries, the centers are rethinking their missions. Some people believe the centers should reduce or even cease cultivar development, in favour of development solely of genetically enhanced breeding materials, to be made available to all plant breeders in developing countries. Further, the centers often are sites for large germplasm depositories, and some believe the centers thus uniquely are suited to characterize and choose potentially useful materials in their collections of landraces and/or wild related species, and then to pre-breed—genetically enhance—them for use by others for cultivar development.

These centers have the same problem, potentially, as breeders in U.S. land grant universities and other tax-supported institutions: in the absence of widespread recognition of genetic enhancement as a profession with easily identifiable products whose value can be calculated, centers that concentrate on genetic enhancement might lose their popular recognition and thus eventually the public funds that support them. This points up further the need for better definition and recognition of the products of genetic enhancement, and of its proper place in plant breeding and variety development. Cultivar development, of course, has had no lack of professional recognition; in fact, as a profession, it has become a successful commercial activity. Many plant breeders now work in private industry, they are supported by the products of their activity-improved cultivars and hybrids-and their numbers increase, annually.

But, plant breeders working for private industry never have done much prebreeding. Private breeders usually depend on the public sector to incorporate needed exotic genes, to remold exotic strains into adapted breeding materials. Genetically enhanced stocks, made by public breeders for their own use in cultivar production, have been freely available to private breeders as well, to use as they wish; privately employed breeders, pressed for short-term results, have depended heavily on publicly employed breeders for genetically enhanced stocks.

IS GENETIC ENHANCEMENT A DYING ART?

And herein arises a problem. Trends are for public breeders to abandon cultivar production and move to more basic biological studies. Beginning young plant breeding scientists in the public sector are going in this more basic direction. Breeders of the older generation are not redirecting their activities, but they are retiring. The gradual change resulting from these two trends is breaking the connection in the chain of development from basic biology through genetic enhancement and on to cultivar production. Publicly employed plant biologists by and large are entering the plant breeding research field at levels well before genetic enhancement-in the more precisely defined (and better funded) science of molecular biology. A gap thus may be developing, in the progression from basic plant biology to final cultivar production.

Some public breeders, seeing this problem, are trying to build up support specifically for genetic enhancement, but because genetic enhancement is not a well-recognized discipline—and perhaps because some people feel the term "genetic enhancement" is only a plant breeder's euphemism for final cultivar production—its support is neither large nor widespread. Private breeders, in general, have not taken up the torch of pre-breeding—or if so, are not talking about it. We may be building towards a shortage of genetically-enhanced breeding materials, a shortage that will become evident a few years further on.

A NEW CONCEPT OF GENETIC ENHANCEMENT

Perhaps I should differentiate among kinds of generic enhancement. Genetic enhancement, up to now, has utilized standard hybridization, segregation, and whole plant selection techniques. Backcrossing, cyclic population improvement, and pedigree selection among selfed progeny are examples of methods employed. But with the advent of molecular genetics and cell biology, a new kind of biotechnology-assisted genetic enhancement (or prebreeding) is possible. Heretofore unavailable genes for insect resistance may be transferred from alien species into elite genotypes (e.g., *Bt* genes from bacteria to maize); or, cellular fusions may be used to construct new systems of cytoplasmic male sterility (e.g., in *Brassica* species); or, restriction fragment length polymorphisms (RFLPs) may be used to build up tables of RFLP linkages with agronomically useful traits (e.g., as in wheat, to allow faster, planned introgression of alien germplasm). Such novel genetic potentials have raised the possibility of entirely new product lines-new kinds of plant cultivars, never before available.

The prospect of being able to market novel breeding stocks or novel varieties, produced via genetic engineering or other new cell and/or DNA technologies, has attracted the attention of entrepreneurially inclined scientists and business people. New companies, or new divisions in existing companies, have been formed to exploit the new commercial opportunities in plant biology.

Commercial applications of biotechnology in plant prebreeding activities generally have been along one of two lines:

1. Some companies have developed fully integrated research and development programmes, designed to utilize a combination of molecular biology and standard plant breeding techniques for prebreeding and final cultivar production. Their intention is to turn out, and market, finished cultivars with novel traits or with greatly improved existing traits.
2. Other companies (or sometimes the same companies) have opted to utilize molecular genetics or cell biology to modify—to genetically enhance—existing elite germplasm stocks which then can be marketed as source materials for development of new cultivars containing the novel, engineered traits. The "engineered genetic enhancement" may be done on contract, or it may be done with expectation that the new enhanced genotypes will be marketed (via sale or licensing) to end-users such as seed companies or integrated food-processors. In either case, genetic enhancement is done as a commercial end in itself; the end-users, the purchasers of the enhanced materials, will do the final step of cultivar development.

A NEW FUNDING BASE FOR GENETIC ENHANCEMENT?

But these biotechnology-using genetic enhancers are different from the earlier generation of genetic enhancers. Not only do they use different technology (and, often, different source materials, such as bacteria), they also have a different source of funding. They do not depend on public funds; they intend to be supported by profits from sale or licensing of their enhanced germplasm. They intend to be paid by the seed companies that use their products. The seed companies in turn are expected to raise the price of their seed as sold to the farmer, to cover the extra costs of cultivar development and thus keep profit margins at safe levels. So in the end, the farmer would be expected to pay for genetic enhancement, whereas up to now the public at large has borne the cost. In some cases, genetically-enhanced materials, produced via new biotechnologies, might be used to reduce a seed company's costs of breeding or production, e.g., if new kinds of male sterile systems for hybrid seed production were to be developed. In such cases price increases for seed would not be required. But in general, genetic engineering is envisaged as adding a desired and affordable extra value—extra profitability—to crops grown by the farmer. The farmer would-according to this plan-be able to afford an extra charge for the seed because of the extra value in the seed.

The new payment scheme for generic enhancement thus opens the door for private enterprise and in theory will bring in funds needed for development of genetic enhancement. The scheme as yet exists only in theory, however. Up to this time, biotechnology-based genetic enhancement businesses have been supported mainly by venture capital funds, not by sale of genetically-enhanced product. Prices, and terms of sale, license or partnership only now are being worked out. And such arrangements necessarily must be made with no knowledge or experience of how much added value really will accrue from the genetically-enhanced materials, or when such value might be ready for sale in products, or whether farmers would want it, or be willing—and able—to pay for it. Thus, the new kind of genetic enhancement-genetic engineering enhancement-still is an unproven accomplishment, financially as well as technically.

If it succeeds, it will add a new dimension to the economics of farming, for farmers will be paying a larger share of the development cost of their seed inputs, and the general public (through general tax funds) will pay less. And a new category of food-and-agriculture-related workers in private enterprise will be formed: genetic engineering prebreeders.

But a question arises: if this new scheme (of using private enterprise for genetic enhancement via biotechnology) succeeds, will traditional genetic enhancement (without aid of biotechnology) still remain as a separate, publicly-supported discipline? Or will those who are practicing genetic enhancement-for-hire with the tools of biotechnology add the tools of standard genetics and plant breeding to their repertoire and proceed to do "all-purpose genetic enhancement," with the goal simply of developing and selling useful genetically-enhanced breeding stocks, produced with whatever technology works best? Such a development might erase the currently popular but really artificial distinction between biotechnology and older technologies, used in aid of genetic enhancement.

Genetic enhancement's financial future is uncertain, as is its technological future. Much will depend on whether some of its products obtained via any technology can be produced in relatively short term, with obvious and quantifiable value, and at a price that buyers can afford. If so, commercial funds and talent will be attracted; if not, commercial efforts at prebreeding as an end in itself will be abandoned. Then public funding, and scientists at public institutions, will need to continue to carry the load, unless the seed industry steps up rates and amounts of in-house activity in genetic enhancement, with or without aid of biotechnology.

Even if genetic enhancement does become a commercially successful enterprise, publicly-funded efforts will continue to be needed for those basic, preliminary kinds of genetic enhancement (early prebreeding) that require long-term effort with uncertain or even unpredictable results. Such long-term, chancy work will not attract private enterprise. And, many seed crops are not dealt with at all by private enterprise breeding, due to their small scale or otherwise low profit margin potential. Thus, for well-rounded genetic enhancement programmes in any crop, public research in genetic enhancement always will be needed.

USES OF BIOTECHNOLOGY

Biotechnology-the sum of the technologies deriving from molecular and cell biology-will have utility in plant breeding beyond the prebreeding stages. Genetic transformation—nonsexual insertion of alien genes and/or gene regulating systems—is envisaged as eventually being applied, quickly and easily, to finished cultivars, to endow them with a series of useful new traits. However, experience is beginning to show that genetic background affects expression of genes and gene systems that are transferred in vitro, as is also the case when they are transferred via regular sexual mechanisms. Long periods of trial and error are needed to bring transformed lines into desired levels of expression. Thus, for some years to come, genetic transformation perhaps will be used only as a prebreeding technique, one in which alien genes are put into breeding pools of elite germplasm, following which standard plant breeding methods are used to extract good new varieties containing the alien genes.

Cell culture techniques—particularly protoplast fusion techniques—will have utility for making new combinations of nuclear and cytoplasmic genes. Immediate utility is seen for manufacturing new male sterility systems to make F1 hybrid cultivars, or to develop new kinds of herbicide resistance. With time and experience it may be possible to increase primary productive potential as well, by selecting and fixing optimum mitochondrial and chloroplast genotypes, balanced with proper nuclear genotypes. Such success seems far in the future, however, since at present little is known about optimization of nuclear, mitochondrial, and chloroplastic combinations. DNA technology will help plant breeding in a more traditional way, when information is further developed on maps of restriction fragment length polymorphisms (RFLPs) and their linkages with useful traits in crop plants. In vitro tests, using RFLP markers, will allow near perfect accuracy in selecting for such useful linked traits. RFLP technology also will be used for precisely characterizing cultivar lineages and relationships, for positive identification of miss-identified or mislabeled cultivars, for characterizing gene frequencies of populations, and even to test hypotheses of landrace evolution, and global movement of crop genotypes.

Perhaps the greatest contribution of molecular biology to plant breeding will be a deeper and more precise understanding of gene action and inheritance, of cellular physiology and biochemistry, and perhaps of whole plant physiology. With such deepened understanding, breeding and selection via standard sexual methods will be planned and executed with greater vision and economy. Such knowledge and execution will be, in part, a consequence of efforts to make genetic transform-nation a practical reality. The fall-out of knowledge from genetic engineering efforts, wisely utilized by observant biologists, will be of great benefit to plant breeding, and to plant culture in general.

DEVELOPMENT OF NEW CROPS

In addition to broadening the genetic base of established crops, genetic enhancement-and plant breeding in the sense of final cultivar development-can be used in two other ways: to develop new crops from heretofore uncultivated species, and to change old crops into new crops.

Breeding of the first type (creating new crops where none existed before) usually is interpreted also to include development of advanced cultivars from landraces. Usually the advanced cultivars are made for industrial countries and the landraces come from developing countries. Soybeans and amaranth thus are spoken of as "new crops" in the United States, even though they have been cultivated for centuries as landraces in China, and in Central and South America, respectively

Challenges are great and are about the same, for changing either wild species or landraces into advanced cultivars, able to withstand monocropping and intensive culture at high production levels. Selection for productivity inevitably requires narrowing the germplasm base, in the course of selecting plant architecture and physiology able to withstand crowding and the stress of high yields, and amenable to machine planting, weeding and harvesting. Planting the narrowly selected genotypes in large areas gives excellent access to specially adapted disease and insect pests, as already noted, and so the need for tolerance or resistance always is next to be added to the list of breeding priorities for new crops. Thus, genetic enhancement first narrows and then later must broaden the genetic base, in the course of new crop development.

And as soon as users of the new product get used to larger quantities of it, they will want to standardize and then to change the quality of the product-as, in oil percentage, grain storability and hardness, or flavor and texture. The rule is simple: the more successful the crop, economically, the greater the demands on (and complaints about) the plant breeders.

Breeding of the second type-changing old crops into new ones-has been considered as a broad-scale possibility only since concepts of genetic engineering were developed. Ability to produce human insulin in bacteria gave rise to thoughts of producing insulin, or other useful primary gene products, in crop plants. Plants were envisaged as now able to turn sunshine, air and water (and soil nutrients) into high value specialty chemicals, instead of commodity chemicals like corn starch or soybean oil.

Most of these ideas still are only ideas, but workers are proceeding to put some of the least complicated ones into action. Modification, via genetic engineering, of the ratios of fatty acid components of rapeseed oil is contemplated. Standard breeding to modify chemical components has been used in Canada already, to make rapeseed into a new crop: "canola." Canola oil, unlike standard rapeseed oil, is essentially free of erucic acid and glucosinolates and therefore is safe for human consumption when used as a cooking oil.

Another possibility derives from recent success in cloning the R gene (first identified by Mendel) in peas which provides the opportunity to identify and transfer, in vitro, genes giving new starch branching properties and thus, potentially, useful new kinds of specialty starches in old starch-producing crops such as maize or wheat. A word of caution is needed for those looking to replace major portions of commodity crop production areas with specialty-product new versions of the old crops. High value specialty products, by definition, require only small amounts of product to saturate the market. They won't replace very much of the production area now used for production of commodity crops. On the other hand, they will make intensive use of crop breeding and crop management skills. They are "brain-power intensive" crops, and as such, must command higher preunit prices—premiums—or they won't be grown.

PHOTOSYNTHESIS

PHOTOLYSIS AND CARBON FIXATION

Photosynthesis is the means that primary producers (mostly plants) can obtain energy via light energy. The energy gained FROM light can be used in various processes mentioned below for the creation of energy that the plant will need to survive and grow.

Photosynthesis is a reduction process, where hydrogen is reduced by a coenzyme. This is in contrast to respiration where glucose is oxidised.

The process is split INTO two DISTINCT areas, *photolysis* (the photochemical stage) and the *Calvin Cycle*. The diagram below gives a summary of the reaction, where light energy is used to initiate the reaction in its presence;

$$CO_2 + H_2O \rightarrow \text{glucose} + \text{oxygen}$$

Photolysis

This part of photosynthesis occurs in the *granum* of *a chloroplast* where light is absorbed by *chlorophyll*; a type of photosynthetic pigment that converts the light to chemical energy. This reacts with water (H_2O) and splits the oxygen and hydrogen molecules apart.

From this dissection of water, the oxygen is released as a by-product while the reduced hydrogen acceptor makes its way to the second stage of photosynthesis, the Calvin cycle.

Overall, since the water is oxidised (hydrogen is removed) and energy is gained in photolysis which is required in the Calvin cycle

The Calvin Cycle

Also known as the carbon fixation stage, this part of the photosynthetic process occurs in the *stroma* of chloroplasts. The carbon made available FROM breathing in carbon dioxide enters this cycle, which is illustrated below:

Just like the *Kreb's Cycle* in respiration, a substrate is manipulated INTO various carbon compounds to produce energy. In the case of photosynthesis, the following steps occur, which create glucose for respiration FROM the carbon dioxide introduced INTO the cycle;

- Carbon FROM CO_2 enters the cycle combining with Ribulose Biphosphate (RuBP)
- A compound formed is unstable and breaks down FROM its 6 carbon nature to a 3 carbon compound called glycerate phosphate (GP)
- Energy is used to break down GP INTO triose phosphate, while a hydrogen acceptor reduces the compound therefore requiring energy
- Triose Phosphate is the end product of this, a 3 carbon compound which can double up to form glucose, which can be used in respiration.
- The cycle is completed when the leftover GP molecules are met with a carbon acceptor and then turned INTO RuBP, which is to be joined with the carbon dioxide molecules to re-begin the process.

The energy that is used up in the Calvin cycle is the energy that is made available during photolysis. The glucose that is made via GP can be used in respiration or a building block in forming *starch* and *cellulose,* materials that are commonly in demand in plants.

Limiting Factors in Photosynthesis

Some factors affect the rate of photosynthesis in plants, as follows

- *Temperature* plays a role in affecting the rate of photosynthesis. Enzymes involved in the photosynthetic process are directly affected by the temperature of the organism and its environment
- Light Intensity is also a *limiting factor,* if there is no sunlight, then the photolysis of water cannot occur without the light energy required.
- Carbon Dioxide concentration also plays a factor, due to the supplies of carbon dioxide required in the Calvin cycle stage.

Overall, this is how a plant produces energy which supplies a rich source of glucose for respiration and the building blocks for more complex materials. While animals get their energy FROM food, plants get their energy FROM the sun.

METHODS OF PLANT BREEDING

CONVENTIONAL METHODS

Plant breeding is defined as identifying and selecting desirable traits in plants and combining these into one individual plant. Since 1900, Mendel's laws of genetics provided the scientific basis for plant breeding. As all traits of a plant are controlled by genes located on chromosomes, conventional plant breeding can be considered as the manipulation of the combination of chromosomes. In general, there are three main procedures to manipulate plant chromosome combination. First, plants of a given population which show desired traits can be selected and used for further breeding and cultivation, a process called selection. Second, desired traits found in different plant lines can be combined together to obtain plants which exhibit both traits simultaneously, a method termed hybridization. Heterosis, a phenomenon of increased vigor, is obtained by hybridization of inbred lines. Third, polyploidy can contribute to crop improvement. Finally, new genetic variability can be introduced through spontaneous or artificially induced mutations.

Selection

Selection is the most ancient and basic procedure in plant breeding. It generally involves three distinct steps. First, a large number of selections are made from the genetically variable original population. Second, progeny rows are grown from the individual plant selections for observational purposes. After obvious elimination, the selections are grown over several years to permit observations of performance under different environmental conditions for making further eliminations. Finally, the selected and inbred lines are compared to existing commercial varieties in their yielding performance and other aspects of agronomic importance.

Hybridization

The most frequently employed plant breeding technique is hybridization. The aim of hybridization is to bring together desired traits found in different plant lines into one plant line via cross- pollination. The first step is to generate homozygous inbred lines. This is normally done by using self-pollinating plants where pollen from male flowers pollinates female flowers from the same plants.

Once a pure line is generated, it is outcrossed, i. e. combined with another inbred line. Then the resulting progeny is selected for combination of the desired traits. If a trait from a wild relative of a crop species, e.g. resistance against a disease, is to be brought into the genome of the crop, a large quantity of undesired traits (like low yield, bad taste, low nutritional value) are transferred to the crop as well. These unfavourable traits must be removed by time-consuming back-crossing, i. e. repeated crossing with the crop parent. There are two types of hybrid plants: interspecific and intergeneric hybrids. Beyond this biological boundary, hybridization cannot be accomplished due to sexual incompatibility, which limits the possibilities of introducing desired traits into crop Plants.Heterosis is an effect which is achieved by crossing highly inbred lines of crop plants. Inbreeding of most crops leads to a strong reduction of vigor and size in the first generations.

After six or seven generations, no further reduction in vigor or size is found. When such highly inbred plants are crossed with other inbred varieties, very vigorous, large sized, large-fruited plants may result. The term "heterosis" is used to describe the phenomenon of hybrid vigor. The most notable and successful hybrid plant ever produced is the hybrid maize. By 1919, the first commercial hybrid maize was available in the United States. Two decades later, nearly all maize was hybrid, as it is today, although the farmers must buy new hybrid seed every year, because the heterosis effect is lost in the first generation after hybridization of the inbred parental lines.

Polyploidy

Most plants are diploid. Plants with three or more complete sets of chromosomes are common and are referred to as polyploids. The increase of chromosomes sets per cell can be artificially induced by applying the chemical colchicine, which leads to a doubling of the chromosome number. Generally, the main effect of polyploidy is increase in size and genetic variability. On the other hand, polyploid plants often have a lower fertility and grow more slowly.

Induced mutation

Instead of relying only on the introduction of genetic variability from the wild species gene pool or from other cultivars, an alternative is the introduction of mutations induced by chemicals or radiation. The mutants obtained are tested and further selected for desired traits. The site of the mutation cannot be controlled when chemicals or radiation are used as agents of mutagenesis. Because the great majority of mutants carry undesirable traits, this method has not been widely used in breeding programmes.

Success in using conventional plant breeding principles and agricultural techniques reached its peak when high-yielding wheat and rice lines were cultivated in the 1960s. The doubling and tripling of productivity of these important crops in Asia signaled a agricultural revolution in the developing countries. This breakthrough in food production was termed the "Green Revolution" to describe the social, economic, and nutritional impact of the new high-yielding wheat and rice strains. Norman Borlaug was awarded the Noble Price in 1970 for his contribution in breeding new high-yielding strains of cereals. However, these strains were highly dependent on fertilizers, irrigation and agrochemicals and required energy-intensive investments. This led to degradation and loss of soils and other severe environmental problems all over the world. For these reasons, the "Green Revolution" has been both praised and damned.

BIOTECHNOLOGICAL METHODS

Biotechnology is the discipline which deals with the use of living organisms or their products. In this wide sense, also traditional agriculture may be seen as a form of biotechnology. The European Federation of Biotechnology defines biotechnology as "the integrated use of biochemistry, microbiology and engineering sciences in order to achieve technological (industrial) application of the capability of microorganisms, cultured tissue cells and parts thereof". In recent years, biotechnology has developed rapidly as a practical means for accelerating success in plant breeding and improving economically important crops.

In Vitro Cultivation of Plant Cells and Regeneration of Plants from Cultured Cells Certain isolated somatic plant cells can be cultured in vitro (in the test tube) and are capable of proliferation and organization into tissues and eventually into complete plants. The process of regenerating whole plants out of plant cells is called in vitro regeneration. The three factors affecting plant regeneration are genotype, explant source, and culture conditions, including culture medium and environment. Different mixtures of plant hormones and other compounds in varying concentrations are used to achieve regeneration of plants from cultured cells and tissues. As the plant hormonal mechanisms are not yet understood completely, the development of in vitro cultivation and regeneration systems is still largely based on empirically testing variations of the three above mentioned factors.

In Vitro Selection and Somaclonal Variation: Plants regenerated from cell cultures may exhibit phenotypes differing from their parent plants, sometimes at quite high frequencies. If these are heritable and affecting desirable agronomic traits, such "somaclonal variation" can be incorporated into breeding programmes. However, finding of specific valuable traits by this method is largely left to chance and hence inefficient. Rather than relying on this undirected process, selection in vitro targets specific traits by subjecting large populations of cultured cells to the action of a selective agent in the Petri dish. For purposes of disease resistance, this selection can be provided by pathogens, or isolated pathotoxins that are known to have a role in pathogenesis. The selection will only allow those cells to survive and proliferate that are resistant to the challenge. Selection of cells also plays an important role in genetic engineering, where special marker genes are used to select for transgenic cells.

Somatic Hybrid Plants: Somatic hybrid plants are plants derived from the fusion of somatic cells. Cell fusion was developed after the successful culture of a large number of plant cells stripped of their cell walls. The resulting cells without walls are referred to as protoplasts. Since also protoplasts from phylogenetically unrelated species can be fused, attempts have been made to overcome sexual incompatibility using protoplast fusion. In most cases, these attempts failed because growth and division of the fused cells did not take place when only distantly related cells were fused. Successful fusions between sexually incompatible petunia species and between potatoes and tomatoes did not lead to economically interesting products, but important contributions to the understanding of cell wall regeneration and other mechanisms were achieved.

GENETIC ENGINEERING

Genetic engineering is a term used for the directed manipulation of genes, i. e. the transfer of genes between organisms or changes in the sequence of a gene. Closely related to this field are methods which use genes or specific sequences for the identification of traits and other analytical purposes. In plant breeding, the most important and already widely used method of this kind is Restriction Fragment Length Polymorphism (RFLP).

Restriction Fragment Length Polymorphism (RFLP)

The techniques of traditional breeding are very time-consuming. By making crosses, also a large number of undesired genes is introduced into the genome of the plant. The undesired genes have to be "sorted out" by back-crossing. The use of Restriction Fragment Length Polymorphism greatly facilitates conventional plant breeding, because one can progress through a breeding programme much faster, with smaller populations and without relying entirely on testing for the desired phenotype.

RFLP makes use of restriction endonucleases. These are enzymes which recognize and cut specific nucleotide sequences in DNA. The sequence GAATTC is cut by the endonuclease EcoRl. After treatment of a plant genome which restriction endonucleases, the plant DNA is cut into pieces of different length, depending on the number of recognition sites on the DNA. These fragments can be separated according to their size by using gel electrophoresis and are made visible as bands on the gel by hybridizing the plant DNA fragments with radiolabeled or fluorescent DNA probes. As two genomes are not identical even within a given species due to mutations, the number of restriction sites and therefore the length and numbers of DNA fragments differ, resulting in a different banding pattern on the electrophoresis gel.

This variability has been termed restriction fragment length polymorphism (RFLP). The closer two organisms are related, the more the pattern of bands overlap. If a restriction site lies close to or even within an important gene, the existence of a particular band correlates with the particular trait of a plant, e.g. disease resistance. By looking at the banding pattern, breeders can identify individuals which have inherited resistance genes, and resistant plants can be selected for further breeding. The use of this technique will not only accelerate progress in plant breeding considerably, but will also facilitate the identification of resistance genes, thereby opening new possibilities in plant breeding.

Gene Transfer

In conventional breeding, the pool of available genes and the traits they code for is limited due to sexual incompatibility to other lines of the crop in question and to their wild relatives. This restriction can be overcome by using the methods of genetic engineering, which in principle allow introducing valuable traits coded for by specific genes of any organism (other plants, bacteria, fungi, animals, viruses) into the genome of any plant. The first gene transfer experiments with plants took p[illegible] 1980s. Normally, transgenes are inserted into the nuclear genome of a p[illegible] cell. Recently it has become possible to introduce genes into the ge[illegible] of chloroplasts and other plastids (small organelles of plant [illegible] which possess a separate genome).

Transgenic plants have been obtained using Agrobacterium-mediated DNA-transfer and direct DNA-transfer, the latter including methods such as particle bombardment, electroporation and polyethylenglycol permeabilisation. The majority of plants have been transformed using Agrobacterium mediated transformation.

Agrobacterium-mediated Gene Transfer

The Agrobacterium-mediated technique involves the natural gene transfer system resident in the bacterial plant pathogens of the genus Agrobacterium. In nature, Agrobacterium tumefaciens and Agrobacterium rhizogenes are the causative agents of the crown gall and the hairy root diseases, respectively. The utility of Agrobacterium as a gene transfer system was first recognized when it was demonstrated that these plant diseases were actually produced as a result of the transfer and integration of genes from the bacteria into the genome of the plant. Both Agrobacterium species carry a large plasmid (small circular DNA molecule) called Ti in A. tumefaciens and Ri in A. rhizogenes.

A segment of this plasmid, designated T-(for transfer) DNA, is transmitted by this organism into individual plant cells, usually within wounded tissue. The T-DNA segment penetrates the plant cell nucleus and integrates randomly into the genome where it is stably incorporated and inherited like any other plant gene in a predictable, dominant Mendelian fashion. Expression of the natural genes on the T-DNA results in the synthesis of gene products that direct the observed morphological changes, i. e. tumor or hairy root formation.

In genetic engineering, the tumor inducing genes within the T-DNA which cause the plant disease are removed and replaced by foreign genes. These gen[illegible]ly integrated into the genom[illegible] plant after infection with the altered strai[illegible]grobacterium, just like the natural T-DNA. Be[illegible] all tumor-inducing genes are removed, the gene t[illegible]er does not induce any disease symptoms. This reliable method of gene transfer is well suited for plants which are susceptible to infection by Agrobacterium. Unfortunately, many species, especially economically important legumes and monocotyledons such as cereals, do not respond positively to

Agrobacterium-mediated transformation. For these plants, the following methods of direct DNA uptake must be applied..

Particle Bombardment

This method, also referred to as biolistic transformation (from biological ballistics) involves coating biologically active DNA onto small tungsten or gold particles (1-5 ?m in diameter) and accelerating them into plant tissue at high velocity. The particles penetrate the plant cell wall and lodge themselves within the cell where the DNA is liberated resulting in transformation of the individual plant cell in an explant. This technique is generally less efficient than Agrobacterium-mediated transformation, but has nevertheless been particularly useful in several plant species, most notably in cereal crops. The introduction of DNA into organized, morphogenic tissues such as seeds, embryos or meristems has enabled the successful transformation and regeneration of rice, wheat, soybean and maize, thus demonstrating the enormous potential of this method.

Electroporation and Direct DNA Entry into Protoplasts

Electroporation is a process whereby very short pulses of electricity are used to reversibly permeabilize lipid bilayers of plant cell membranes. The electrical discharge enables the diffusion of macromolecules such as DNA through an otherwise impermable plasma membrane. Because the plant cell wall will not allow the efficient diffusion of many transgene constructs, protoplasts (cells without cell walls) must be prepared. This requirement presents a major obstacle for many applications as protocols making possible the regeneration of protoplasts into complete plants do not exist for many species.

Gene transfer from Agrobacterium tumefaciens to a plant cell. In nature (route A), the transfer DNA (T-DNA) contains tumor inncluding genes which lead to crown gall disease. In plant genetic engineering (route B), the tumor including genes are removed and replaced by other genes, e.g. genes which confer insect resistance. In nature as well as in genetic engineering, the agrobacteria attach to a plant cell. Then, the T-DNA is cut out of a plasmid (a small circular DNA-molecule) and is transfered to the plant cell. The T-DNA migrates to the nucleus of the plant cell and becomes incorporated into a plant chromosome. In nature (route A), the tumor inducing genes cause uncontrolled (tumor-like) growth at the site of infection after expression of the T-DNA. In genetic engineering (route B, tumor inducing genes removed by other genes), transgenic cells are selected and regenerated to whole plants. All cells of the regenerated plant now contain the transgene. Adapted from

DNA uptake by plant protoplasts can also be stimulated by phosphate or calcium/polyethylene glycol (PEG) coprecipitation. However, these methods all suffer from the drawback that they use protoplasts as the recipient host which often cannot be regenerated into whole plants.

Transgene Expression

In most cases, the introduction of a gene into the plant genome will only have an effect on the plant if the transgene is expressed, i. e. transcribed into mRNA and translated into a protein. A promoter is a sequence of nucleic acids where the RNA polymerase (a complex enzyme synthesizing the mRNA transcript) attaches to the DNA template. The nature of the promoter defines (together with other expression-regulating elements), under which conditions and with which intensity a gene will be transcribed. The promoter of the 35S gene of cauliflower mosaic virus is used very frequently in plant genetic engineering. This promoter confers high-level expression of exogenous genes in most cell types from virtually all species tested. As it is often advantageous to express a transgene only in certain tissues or quantities or at certain times, a number of other promoters are available, e.g. promoters inducing gene expression after wounding or during fruit ripening only.

Methods of gene transfer currently employed result in the random integration of foreign DNA throughout the genome of recipient cells. The site of insertion may have a strong influence on the expression levels of the exogenous gene, resulting in different expression levels of an introduced gene, even if the same promoter/gene construct was used. The exact mechanism of this phenomenon are not yet fully understood.

Selection and Plant Regeneration

In a transformation experiment, the proportion of transformed cells is usually small compared to the number of cells which remain unaltered. In order to select only cells which have actually incorporated the new genes, the genes coding for the desired trait are fused to a gene which allows selection of transformed cells, so-called marker genes. The expression of the marker gene enables the transgenic cells to grow in presence of a selective agent, usually an antibiotic or a herbicide, while cells without the marker gene die. One of the most commonly used marker genes is the bacterial aminoglycoside-3' phosphotransferase gene (APH(3')II), also referred to as neomycin phosphotransferase 11 (NPTII). This gene codes for an enzyme which inactivates the antibiotics kanamycin, neomycin and G418 through phosphorylation. In addition to NPTII, a number of other antibiotic resistance genes have been used as selective markers, e.g. hygromycin phosphotransferase gene conferring resistance to hygromycin.

Another group of selective markers are herbicide tolerance genes. Herbicide tolerance has been obtained through the incorporation and expression of a gene which either detoxifies the herbicide in a similar manner as the antibiotic resistance gene products or a gene that expresses a product which acts like the herbicide target but is not affected by the herbicide. Herbicide tolerance may not only serve as a trait useful for selection in the development of transgenic plants, but also has some commercial interest. Herbicide tolerance transgenic plants are therefore among the first crops approaching market introduction.

Transformation of plant protoplasts, cells and tissues is usually only useful if the they can be regenerated into whole plants. The rates of regeneration vary greatly not only among different species, but also between cultivars of the same species. As mentioned in capter 2.2, in many cases regeneration of whole plants from cells is not possible or very difficult. Besides the ability to introduce a gene into the genome of a plant species, regeneration of intact, fertile plants out of transformed cells or tissues is the most limiting step in developing transgenic plants.

STATE OF THE ART GENETIC ENGINEERING FOR PLANT PROTECTION

RESISTANCE OF TRANSGENIC PLANTS

Strategies using genetic engineering to achieve disease and pest resistant plants are advancing rapidly. Different degrees of resistance against insects, viruses, fungi and bacteria have been reached with various crop species.

Resistance to Plant-Feeding Insects

World-wide losses resulting from plant-feeding insects are still more than 10 percent of total production, despite the intensive use of a large number of chemical insecticides. The use of these agrochemicals is not only an important economical factor in agriculture, with annual expenses of more than $7.8 billion world-wide, but has also often resulted in ecological problems. In addition, more than 500 target insect species have developed resistance to chemical insecticides. For these reasons, several approaches of genetically engineered insect resistance have been developed.

The use of Bacillus thuringiensis toxins has by far been the most widely used and successful strategy, and several Bt expessing crops have obtained marketing clearance. Some other approaches for insect resistance in transgenic plants are in development and may lead to successful insect resistance in the near future.

Bacillus Thuringiensis ä-Endotoxins

Biological Basis

Bacillus thuringiensis (Bt) is a common soil bacterium which forms resistant resting structures under adverse conditions, so called spores. During formation of these spores, Bt produces a crystal which is predominantly comprised of one or more proteins called ä-endotoxins, Bt toxins or insecticidal crystal proteins. These ä-endotoxins possess insecticidal activity when ingested by certain insects. Intensive investigations have led to detailed knowledge of the mechanism and specificity of Bt toxin activity. Upon ingestion by susceptible insect larvae, the crystalline inclusions are solubilized in the midgut, releasing one or more proteins of different sizes. Rapid proteolytic cleavage of these polypeptides then produces the active toxic proteins. These activated toxins bind to specific high-affinity receptors on the insect midgut cell membranes, leading to the formation of pores that disturb the cellular osmotic balance. Within minutes, midgut cells are paralyzed and disrupted.

Bacillus thuringiensis bacteria have been used since 1938 to produce an insecticidal spray against certain insect pests. These products were, and still are, produced by fermentation of single Bt strains in crude, inexpensive media. Numerous strains of Bt are currently known. Each strain produces differing numbers of ä-endotoxins with various insecticidal activities. A given ä-endotoxin is typically insecticidal only towards a narrow spectrum of insect targets. Screening efforts have yielded strains insecticidal to Lepidoptera, Diptera, Coleoptera, and even some strains effective against nematodes. Although Bt bacteria are an effective microbial pesticide, a number of biological constraints limit their use: The Bt ä-endotoxins are short lived when sprayed on crops, thereby reducing the residual activity and necessitating many applications during a growing season. Bt, as with other spray-on insecticides, is difficult to deliver to insect species which burrow into their host plant, hide under leaves or live primarily under the soil surface.

These limitations can be overcome by expressing Bt toxins in transgenic plants. Since Bt toxin is the product of a single gene, and its safe use and efficacy has a long history, the ä-endotoxin genes were among the first genes of commercial interest to be engineered into plants.

In the beginning, plants were transformed with full length Bt toxin genes. These plants mostly showed only low levels of toxin mRNA and protein and did not exert significant insecticidal activity. Better results were obtained using truncated versions of Bt genes, but in many cases the level of Bt toxin expression and therefore protection was still to low to be of agronomic relevance.

Significant increases were achieved by modification of the truncated structural gene that had either none or only a minor effect on the encoded amino acid sequence.[9] Alterations affected 5'-or 3'- terminal regulatory mRNA sequences, mRNA secondary structure, G+C content and/or codon usage. A truncated crylA(b) gene encoding amino acids 1-648 of crylA(b) gene from Bacillus thuringiensis var. kurstaki HD-1.

This synthetic gene was completely redesigned to replace the bacterial codons with maize- preferred codons. The synthetic gene was combined with different promoters and brought into maize. Field trials showed that the transgenic maize lines provided season-long protection from repeated heavy infestations of European corn borer, which totally devastated control plants.

Although the redesign of the ä-endotoxin genes is a successful means to enhance the Bt toxin production in plants, it suffers from the disadvantage of being labourious and expensive. A promising new solution for enhanced translation efficiency of Bt toxin genes was reported. In this work, the transgene was not introduced into the nucleus, but into plastids, a group of small organelles of plant cells including chloroplasts with an own genome called the plastone. The plastids are thought to be derived from endosymbiontic bacteria and therefore have a transcriptional and translational machinery of procaryotic origin. An unmodified Bt toxin gene was incorporated into the tobacco plastid genome by plastid transformation. Since a plant cell may contain up to 50'000 copies of a plastid, the ä-endotoxin gene was present in several thousands of copies per cell in transgenic plants.

The plastid transformation resulted in an accumulation of an unprecedented 3-5 percent of soluble protein of tobacco cells as Bt toxin and an extremely high toxicity to the insects Heliothis virescens, Helicoverpa zea and Spodoptera exigua. Apart from the high expression level of unaltered Bt genes, the new method of plastid transformation has the additional advantages that position effects can be avoided because the site of introduction of the transgene can be targeted and the transgenes contained in the plastids are not transmitted to other plants by pollen.

As already mentioned, transgenic maize, potato and cotton expressing Bt toxins are among the first transgenic plants arriving on the market in the USA. During the last years, hundreds of field trials in several countries have proven the efficacy of this approach in controlling important insect pests. At present, the majority of transgenic crops containing Bt genes have been transformed with cryI genes which have activity against the lepidopterans, including the tobacco hornworm, tobacco budworm, tomato pinworm, corn earworm and European corn borer.

More recently, the cryIIIA gene has been engineered into plants for protection against coleopterans, including the Colorado potato beetle. Bt ä-endotoxin genes with activity against insects of the genus Diabrotica, such as corn rootworm, a major maize pest, will likely be produced in the next generation of transgenic plants.

Inhibition of Insect Digestive Enzymes

Biological Basis

Many plant species protect themselves against plant-feeding insects by producing proteins which interfere with insect digestion. Proteases are enzymes which brake down proteins in the insect digestive system. Protease inhibitors block the action of these proteases. Most plant protease inhibitors do not effect endogenous plant proteases, but specifically inhibit animal and microbial protease enzymes. This suggests they may be involved in the protection of vulnerable plant tissues from pest and pathogen attack by antinutritional interaction with digestive enzymes. Protease inhibitors are widely distributed within the plant kingdom and can accumulate to particularly high levels in seeds and storage organs. The detrimental effect of protease inhibitors was proven by feeding insects an artificial diet. Rapid systemic accumulation of protease inhibitor mRNA and _proteins was demonstrated in tomato and potato in response to insect attack or mechanical wounding. Another enzyme interfering with insect digestion is á-amylase inhibitor. This inhibitor blocks the action of á-amylase, an enzyme used by insects to degrade starch.

State of Development

The first protease inhibitor protein expressed in transgenic plants was the cowpea trypsin inhibitor (CpTI). A full length CpTI DNA sequence was brought under control of the CaMV 35S promoter and transferred into tobacco. Bioassays considering insect biomass and survival and percentage of leaf area have conclusively established that CpTI-expressing plants have significant resistance to Heliothis virescens and to a broad range of lepidopteran pests feeding on tobacco. Recent field trials with transgenic tobacco expressing CpTI demonstrated that CpTI had a negative effect on both larval survival and plant damage, but the effect was not always significant. The tomato protease inhibitors PH and the potato proteinase inhibitor PPI II represent two other proteinase inhibitors which have been introduced into plants. They adversely affected some insect species like Manduca sexta (tobacco hornworm, lepidoptera) and Chrysodeixis eriosoma (green looper, lepidoptera). In contrast, no inhibition of feeding was achieved with transgenic tobacco producing PI-I inhibitor from tomato or soybean Kunitz trypsin inhibitor (SBTI), indicating that not every protease inhibitor can be employed for control of a specific pest species.

Two recent developments made possible the breeding of peas resistant to bruchid beetles, important pests of seeds in post harvest storage, by using á-amylase inhibitor. First, it was shown that the growth of two bruchid species is inhibited by low doses of bean áAI, and second, a system for genetic transformation of garden pea became available. The coding sequence of the bean áAI-gene was attached to a seed-specific promoter and the chimeric gene was introduced into garden peas (Pisum sativum L). The seeds of the resulting transgenic peas showed a high degree of resistance to the bruchids cowpea weevil and Azuki bean weevil. This achievement may be particularly important in developing countries, not only because of the large amount of post-harvest losses, but also because of the importance of legumes such as peas as a source of protein.

Lectins

Lectins are proteins capable of binding to carbohydrate moieties of complex carbohydrates without altering the structure of the carbohydrate. They were discovered about 100 years ago when the observation was made that extracts of beans agglutinate red blood cells. Further research showed that many plants contain lectins. They are predominantly found in seeds where they can comprise up to two percent of the total protein. The actual function of lectins in plants is not clearly understood, but they are believed to function as a defence mechanism against a variety of fungal and bacterial pathogens. Deleterious effects were found e.g. against weevil larvae and European corn borer. The mechanism of toxicity is unknown but probably involves the binding of lectin to insect midgut cells and disrupting of cell functions or inhibition of nutrient uptake.

The lectin wheat germ agglutinin has been introduced into maize to achieve insect resistance and field trials have been carried out. Hilder et al. transformed tobacco with a snowdrop (Galanthus nivalis) lectin. They reported enhanced resistance of the transgenic tobacco against the sap-sucking peach potato aphid (Myzus persicae). However, no commercially acceptable level of protection was reached. Resistance against insects feeding on phloem sap may be enhanced by using phloem-specific promoters such as the rice sucrose synthase promoter.

Protection against Viral Infections

Viruses cannot be controlled by chemical means. As viruses can cause substantial losses in many crops, they are indirectly controlled by spraying chemicals against virus vectors (mainly insects), by using certified virus-free material, or by eradicating infected plants. As these measures are not overly successful, the engineering of virus resistance into plants is expected to provide a direct and more efficient control.

Viruses are genetic elements much smaller than e.g. bacterial cells. They do not have an own matabolism and multiply only inside living host cells using the cell metabolic machinery. Viruses consist of either a DNA or RNA genome surrounded by a coat protein, and occasionally other material. When a virus multiplies, the viral genome is released from the coat into the host cell, where the host metabolism starts produtcion of viral proteins and replication of viral genes, finally resulting in the release of new viruses. About 80percent of pathogenic plant viruses have genomes of plus strand single-stranded RNA which may be directly used as mRNA, but also single- or double-stranded DNA viruses cause significatn losses to crops.

Most genetic engineering strategies against viruses are based on the introduction of a virus-derived gene sequence into the genome of the host plant. The products of these sequences (mRNA and proteins) interfere with specific stages in the viral infection cycle, such as virus replication or spread, thereby resulting in a virus resistat plant. Resistance mediated by expression of a viral coat protein is the most advanced approach and has already led to resistant varieties which are expected on the market soon. Some other strategies also confer a high degree of resistance, e.g. replicase-mediated protection, but no commercial products have been developed ye using these methods.

Coat Protein-Mediated Protection

Biological Basis

The genome of plant viruses encodes a coat protein that protects the viral nucleic acid during the transfer from plant to plant and can determine the specificity of the vector. In coat protein-mediated protection (CPMP), a gene for a viral coat protein is introduced into the genome of the plant. The concept of CPMP has been derived from the observation of cross protection: by preinfection with a mild symptomless virus strain, plants are protected against a related but severely damaging strain. The Tobacco Mosaic Virus (TMV) coat protein gene was introduced into the genome of tobacco plants. Transgenic plants expressing high levels of coat protein showed a delay in disease development upon infection with TMV.

Typically, CPMP is achieved by transforming plants with a functional sense coat protein gene under control of a Cauliflower Mosaic Virus (CaMV) 35S promoter that mediates strong constitutive transcription in many cell types. CPMP works against the virus strain from which the coat protein gene originates and against related viruses. The mechanism of this resistance is not fully understood. The degree of resistance was often correlated with the amount of intact coat protein expressed in transgenic plants and resistance could be overcome by high concentrations of virus inoculum. Several lines of evidence suggest that the coat protein produced. by the plant impedes the uncoating of the viral RNA at the beginning of the infection cycle. In some cases, resistance was found in plants expressing only low levels of coat protein, in plants expressing defective coat proteins, or in plants expressing only coat protein mRNA which was not translated into protein. In these examples, resistance may be mediated by the coat protein mRNA.

So far, CPMP is the only method for genetically engineered resistance which has the potential for agronomic use. CPMP has been demonstrated for more than 20 RNA viruses, but no effect on DNA viruses has yet been reported. Tobacco has been used as a model system to study CPMP, but resistance has now also been engineered into important field crops such as potato, tomato, squash, alfalfa, and has also been extended to cereals.

The efficiency of CPMP has been tested in field trials with several plant species. The first field trial was conducted in 1987 with transgenic tomato plants. After mechanical inoculation with tobacco mosaic virus (TMV), at the time of fruit harvest only 5 percent of the transgenic plants expressing the TMV coat protein had developed disease symptoms, compared to 99percent of the control plants. The fruit yield was equal to that of non-infected plants. Other trials have shown similarly high degrees of resistance, demonstrating the usefulness of this approach. The US seed company Asgrow has developed a transgenic squash variety resistant to zucchini yellow mosaic virus and watermelon virus II using coat protein genes of the two viruses.

Abbreviations: AIMV alfalfa mosaic virus, ArMV arabis mosaic virus, BNYVV beet necrotic yellow vein virus, BYDV bailey yellow dwarf virus, BYMV bean yellow mosaic virus, CMV cucumber mosaic virus, MDMV maize dwarf mosaic virus, PLRV potato leafroll virus, PRSV potato ringspot virus, PRV papaya ringspot virus, PVS potato virus S, PVX potato virus X, PVY potato virus Y, RSV rice stripe virus, SMV soybean mosaic virus, TEV tobacco etch virus, TMV tobacco mosaic virus, TRV tobacco rattle virus, TSV tobacco streak virus, TSWV tomato spotted wilt virus, TYLCV tomato yellow leaf curl virus, WMV2 watermelon mosaic virus II, ZYMV zucchini yellow mosaic virus.

Replicase-Mediated Protection

As an alternative to the use of genes encoding structural proteins like coat protein genes, the introduction of sequences coding for non- structural viral proteins has been exploited to create virus resistant plants. There are several reports now describing how expression of a viral RNA-polymerase in transgenic plants leads to resistance. The viral RNA-dependent RNA-polymerase, also called replicase, is used by the virus to generate new RNA copies of the viral RNA genome. When it was discovered that transgenic tobacco plants expressing part of the replicase gene of a TW were highly resistant to this virus and a closely related strain.

The mechanisms conferring resistance are not clear. Resistance was brought about both by replicase genes which coded for functional proteins and by mutated genes which lead to production of non-functional protein or no protein at all. One explanation of resistance is based on the observation that overproduction of one component of the transcription complex (in this case the replicase) can lead to inhibition of transcription. If defective replicase is expressed, resistance may result from the effect of the dysfunctional protein which dominates the functional viral proteins. The efficacy of both proposed mechanisms should increase with enhanced expression of the transgene. However, these explanations are in contrast with several experiments where the highest level of resistance has been observed in the plants with the lowest expression rate of the transgene, an effect which has also been found in coat protein-mediated resistance. In these cases, the resistance may be mediated by mRNA,.

In the last five years, at least eight examples of replicase-mediated resistance in model plants, including resistance against tobacco mosaic virus, pea early browning virus, cucumber mosaic virus, potato viruses X and Y, and cymbidium ringspot virus. Though apparently no field trials have been carried out, the extreme level of resistance observed in some experiments makes it likely that this approach will be useful in the field. A limitation may be the strain specificity of the effect, which could cause problems in genetically heterogenous viral field isolates.

Antisense RNA-Mediated Resistance

Antisense RNA is RNA complementary to mRNA of a particular gene. Antisense RNAs were initially recognized in bacteria as naturally occurring mechanisms of gene expression. Antisense strategies are increasingly used as a means of down regulating specific genes. The most famous application to date is the FlavrSavr^{TM} tomato with an extended shelf life which was the first transgenic food available on market.

The extended shelf life is achieved by down regulating the polygalacturonase (PG) gene responsible for the softening of the tomato using an antisense gene of the PG gene. Expression of antisense RNA in transgenic plants is achieved by introducing the noncoding strand of a gene together with an appropriate promoter into the plant genome. The mode of action of antisense RNA is not clearly understood. The antisense RNA is thought to form a duplex with the target viral RNA by hybridization. The duplexes formed may stimulate degradation of the mRNA by RNAses or could prevent binding of the ribosome, thereby impeding translation into proteins.

Research with antisense genes used for virus resistance has been predominantly conducted with tobacco and potato. The antisense approach was successful in several cases, e.g. in transgenic Russet Burbank potato against potato leafroll virus (PLRV). In this case, sense and antisense constructs against viral coat protein were equally effective. Examples of successful induction of resistance also include tomato golden mosaic virus (TGMV), a single stranded DNA virus which replicates in the nucleus. In several other instances, the protection was less effective, e.g. in tobacco expressing antisense coat protein genes of PVX.

Satellite RNA-Mediated Disease Attenuation

Satellite RNAs are small extragenomic RNA molecules found in some plant viruses. They cannot replicate individually and depend on a helper virus for replication. They have no sequence similarity with their helper virus, and are thus not directly derived from the helper virus genome. Some satellite RNAs are known to attenuate the symptoms of the helper virus. In China, pepper plants are prophylactically inoculated with a cucumber mosaic virus (CMV) strain which contains satellite RNA.

These plants develop only mild symptoms and are protected against infection with a more harmful CW strain not containing the satellite. Transgenic plants expressing satellite RNA were shown to be protected against the severe effects of their helper virus. As a mechanism for satellite RNA-mediated disease attenuation, competition with the viral genomic RNA for a limiting amount of replicase enzyme has been proposed, but other mechanisms seem to contribute to the disease attenuation as well.

The most widely studied satellite RNAs are those associated with cucumber mosaic virus (CMV). Several groups have shown that transgenic plants expressing CMV satellite RNA are tolerant to CMV. Transgenic plants showed almost no symptoms after infection with CMV. Also tobacco plants expressing tobacco ringspot virus (TobRSV) satellite RNA were tolerant to TobRSV. Few researchers envisage field testing of transgenic plants due to the possibilities of producing deleterious satellite RNA.

As mentioned before, cross protection with a CMV strain attenuated by a natural satellite RNA has been shown in the field, not only at a large scale in China, but also at smaller scale in U.S. and Italy. Large-scale tests of transgenic plants expressing satellite RNA are being carried out in China.

Defective Interfering RNA/DNA Protection

Defective interfering (DI) particles are deletion mutants of genomic viral sequences which depend on their parent virus for replication. DI RNAs have frequently been described in animal virus systems but, so far, only rarely in plant virus systems. Like satellite RNAs, DI particles can attenuate the disease symptoms of their parent virus by interfering with its replication, as found for tomato bushy stunt virus (TBSV). As an alternative to the introduction of naturally occurring DI molecules, transformation with artificially constructed Dis have been described. Deletion mutants of several viral RNAs were shown to reduce the replication of the respective parent viruses.

Ribozyme-Mediated Protection

Ribozymes are RNA molecules catalyzing RNA cleavage reactions. Mostly, the cleavage site consists of a consensus structure, called the "hammerhead" motive. The nucleotide region directing the catalysis of the cleavage reaction could be separated from the region where the cleavage occurs and the recognition of the target could be modified by changing the nucleotide sequence of the regions flanking the cleavage site. Thus, ribozymes could be designed which cleave viral sequences. Transgenic plants could express a ribozyme which cleaves viral nucleotides. Whether ribozymes will be active in stably transformed plants, and by which expression strategy a sufficient amount of ribozyme can be produced to achieve a reduction in target gene expression remains to be investigated.

Resistance to Fungal Pathogens

Plants exhibit natural resistance to fungal attack, and disease is the exception rather than the rule. However, the exceptions can be costly and even devastating. Fungal disease remains one of the major factors limiting crop productivity worldwide, with often huge losses set against massive cash inputs for pesticide treatments, exemplified by an estimated cost to the wheat farmers of US$100 million per annum in Europe alone. The total market for fungicides amounts about US$ 5.6 billion world-wide. Hence, there is great interest in the development of novel strategies for protecting crop plants against fungal diseases, to complement traditional plant breeding efforts and to reduce the expense and possible environmental costs of reliance on conventional agrochemical treatments.

Although several promising approaches exist, no transgenic varieties with fungal resistance are expected to approach market introduction in the near future. Strategies for obtaining plants resistant to fungi are largely based on natural plant responses to pathogen attack.

Natural Plant Disease Resistance

Plants, like animals, are continually exposed to pathogen attack. Plant- pathogenic organisms are diverse and include viruses, bacteria, fungi, nematodes and protozoa. These pathogens usually live within the plant cell or in the intercellular space. Because plants lack a circulatory system and antibodies, they have evolved a defence system that is distinct from the vertebrate immune system. In contrast to animal cells, each plant cell is capable of defending itself by a combination of different defence reactions.

Resistance to a pathogen is often correlated with a "hypersensitive response", induced cell death in the host plant at the site of infection. The hypersensitive response (HR) involves generation of active oxygen species, ion fluxes across membranes, cross-linking and strengthening of plant cell wall, production of antimicrobial compounds (phytoalexins) and induction of pathogenesis- related (PR) proteins such as chitinases and glucanases. Though the mechanisms are unknown, HR is thought to be responsible for the limitation of pathogen growth. The HR is hypothesized to trigger a subsequent response to pathogen attack, referred to as systemic acquired resistance, that acts throughout the plant and not only at the site of infection. Systemic acquired resistance reduces the severity of disease caused by all classes of pathogens, including normally virulent pathogens.

Whether a plant is susceptible or resistant to a specific pathogen often depends on its ability to induce HR. Induction of HR follows recognition of specific signal molecules (elicitors) produced by the pathogen. These elicitors are directly or indirectly coded for by so-called avr (for avirulence) genes of the pathogen. The receptors of the plant cell which recognize the elicitors and (indirectly) initiate the HR are coded for by R (for resistance) genes. Therefore, a plant is only protected against pathogens which produce a specific elicitor that is recognized by a specific plant receptor.

In conventional plant breeding, plants with resistance (R) genes against important pathogens have been selected according to their phenotypic traits and the resistance has been introduced into other varieties by hybridization. Unfortunately, resistance of a new variety conferred by a single resistance gene is often overcome after only a few years by the development of a new pathogen race, and the combination of several resistance genes by conventional breeding is even more time- consuming than introduction of a single R gene.

Today, the growing understanding of the mechanisms leading to plant disease resistance is increasingly used for plant genetic engineering. Recently, the first R genes have been identified and described in plants. These include R genes from Arabidopsis sp. (resistance to the bacterium Pseudomonas syringae), tobacco (resistance to tobacco mosaic virus and to P. syringae), tomato (resistance to the leaf fungal pathogen Cladosporium fulvum) and flax (resistance to a leaf rust fungal race). The identification of such R genes will not only facilitate the understanding of plant disease resistance, but also makes possible the direct introduction of R genes in different plant varieties using genetic engineering. By combining different R genes responsive to the same pathogen in one plant, resistant varieties with a more durable resistance could be produced.

Genes acting at other levels of the plant defence system have already been introduced into plants in order to increase resistance against fungi and other pathogens.

Chitinases and Glucanases

Chitin and â-1,3-glucans are major structural polysaccharides of the cell walls of many fungi. The breakdown of these components by plant endochitinases and â-1,3-glucanases, both belonging to the group of the pathogenesis-related (PR) proteins of plants which are induced upon pathogen attack, is thought to inhibit fungal growth. Several types of endochitinases and â-1,3-glucanases have been described. A combination of a bean endochitinase gene with the promoter of the CaMV 35S gene conferring high constitutive expression in a wide variety of plant cells was introduced into tobacco. For evaluation, the transgenic tobacco was grown in presence of Rhizoctonia solani, a soilborne fungus that infects numerous plant species. Seedling mortality or the loss of root fresh weight was clearly reduced compared to control plants, depending on the amount of bean chitinase expressed. Similar effects were found with other 35S-chitinase constructs, but in other cases this approach did not reveal a substantial increase of resistance to chitin-containing fungi.

Recently, a rice chitinase gene has been introduced into protoplasts of an Indica rice variety using polyethyleneglycol-mediated transformation, and fertile plants have been regenerated. This was one of the first reports of regeneration of the important Indica rice, as these varieties have been particularly difficult to regenerate. Transgenic plants expressing the chitinase showed increased resistance to the fungal sheath blight pathogen Rhizoctonia solani.

The combined action of chitinase and glucanase was evaluated. The rice RCH10 chitinase and the alfalfa AGLU1 glucanase under control of CaMV 35S promoters were introduced separately into tobacco. The transgenic lines expressing one of the two hydrolytic enzymes were then selfed, and the homozygous progeny lines were crossed with each others. The resulting lines were infected with the fungal pathogen Cercospora nicotianae. Evaluation of disease development showed that the combination of chitinase and glucanase gave the best protection, with only weak symptoms after substantial delay.

The finding that combination of genes which alone have only limited effect results in significant protection is promising. Similar results have been found in other experiments. Moreover, combinatorial use of antimicrobial genes may also reduce the breakdown of resistance as a result of pathogen mutation.

Ribosome-inactivating Proteins

Ribosome-inactivating proteins (RIPs) are plant proteins which inhibit protein synthesis in target cells by cleavage of 28 S RNA. RIPs do not inactivate "self' ribosomes, but show activity towards ribosomes from distantly related species including fungi. Some ribosomes are among the most potent natural toxins, the best known of which is ricin. Purified barley RIP inhibits fungal growth in vitro, and its effect is synergistically enhanced by chitinases and â-1,3-glucanases. These characteristics were used to improve resistance to Rhizoctonia solani in transgenic tobacco. A barley RIP was put under control of the potato wun1 promoter which mediates wound- and pathogen-inducible transcription in the epidermis of leaves, stems and roots from tobacco. Primary transformed regenerates and R1 progeny grew more vigorously in soil inoculated with Rhizoctonia solani than non-transformed control plants.

Phytoalexins

In several host/pathogen systems a correlation has been found between the concentration of phytoalexins (antimicrobial compounds synthesized by plants) and resistance to specific pathogens. In Vitis vinifera and Picea sitchensis, stilbenes are involved in protection against fungal challenge. Most plants, tobacco, contain the precursor for the formation of stilbenes but lack the enzyme stilbene synthase. After transfer of a stilbene synthase gene from grapevine to tobacco, transgenic plants expressed the foreign gene in response to treatment with fungal elicitor or infection with the broad range fungal pathogen Botrytis cinerea. This work supports the assumption that at least in some host/pathogen systems the synthesis of phytoalexins plays a crucial role for the defence of the host plant.

Other Antifungal Proteins

A large number of other proteins with inhibitory effects on the growth of fungi in vitro'have been identified from plants. Among these are osmotin, a vacuolar protein from tobacco produced in response to salt stress, and small, basic, cysteine-rich proteins of various origins. Also some microorganisms produce such antifungal peptides. These proteins will be tested - alone or in combination - for their effect against fungi in transgenic plants and at least some of them are likely to enhance resistance in transgenic crop plants.

Hypersensitive Cell Death

In the past, crossing-in of race-specific resistance was frequently applied in breeding programmes to protect novel varieties against fungal and other diseases, as mentioned before. However, this procedure provides protection only against a limited number of pathogen races and large- scale growth of new varieties led to the rapid selection of new virulent races and epidemic spread in monocultures. Of course, the same problems could also arise if transgenic plants carrying race- specific resistance genes from other cultivars or other species were grown in the field. Different strategies to obtain durable resistance to a wide range of pathogens using artificial generation of programmed cell death are being pursued.

A promoter fragment of the potato prp11 gene, which mediates rapid and local transcriptional activation selectively after fungal infection, has been combined with the barnase gene from Bacillus amyloliquefaciens encoding a highly cytotoxic RNase. The resulting chimeric construct was transformed into potato.

Rapid synthesis of this RNase in the vicinity of infection sites should initiate necrosis of host cells during early stages of compatible interactions and, therefore, restrict the growth and propagation of biotrophic pathogenic fungi also in this type of interaction, analogous to the naturally occurring hypersensitive cell death in incompatible interactions. Transgenic plants were simultaneously transformed with a chimeric construct which constitutively produces barstar the specific barnase inhibitor from Bacillus amyloliquefaciens. The authors expect, that only in the close vicinity of infection sites the level of cytotoxic barnase will exceed the inhibitor barstar, resulting in strictly localized cell death after infection.

Systemic Acquired Resistance

A hypersensitive response is often followed by a subsequent response called systemic acquired resistance (SAR). SAR acts throughout the plant and protects it against a wide range of pathogens. Establishment of SAR is correlated with the coordinate expression of several groups of genes, so-called pathogenesis-related (PR) proteins. PR-1a is the most highly expressed of these proteins. The gene coding for PR-1a was brought under control of an enhanced 35S promoter and transformed into tobacco. Transgenic tobacco lines showed constitutive high-level expression of the PR-1a protein. This resulted in significant tolerance to infection by the two fungal pathogens Peronospora tabacina and Phytophthora parasitica var. nicotianae.

Although the function of PR-1a protein and other PR proteins is not known yet, high-level expression of such proteins may be useful in creating crops with increased resistance to fungal and other diseases.

Resistance to Bacterial Pathogens

Bacterial diseases are a persistent problem, with no satisfactory alternative to copper sulfate treatment. As these treatments cause environmental problems, there is a great interest in the development of novel strategies. While scientific progress has been im- pressive, examples of agronomically significant resistance to plant-pathogenic bacteria are still scarce.

Lysozyme-Mediated Resistance

Lysozymes, which catalyze the hydrolytic cleavage of bacterial cell wall murein, have been detected in many plant species. As the biochemical and molecular characterization of plant lysozymes and the corresponding genes is still in its infancy, lysozyme genes from other sources have been used for genetic engineering. The expression of lysozyme genes from hen egg white or bacteriophage T4 was achieved in tobacco and potato. One group found an increased resistance of tuber slices from transgenic potato lines to the pathogenic soil bacterium Erwinia carotovora spp. atroseptica.

Cecropins

Cecropins are a family of proteins which form an important component of the immune response of diverse insects. They exhibit antibacterial activity against several Gram-positive and Gram-negative bacteria in vitro, apparently by forming ion channels in the bacterial membrane leading to leakage of cell components and ultimately cell death.

A gene coding for cecropin B from the giant silkmoth Hyalophora cecropia was introduced into tobacco Although considerable amounts of cecropin-mRNA were present in transgenic plants, no cecropin B protein could be detected. As rapid degradation of cecropin B was shown in crude plant extracts, the undetectable level of cecropin B in transgenic tobacco is thought to result from degradation of the protein by plant-endogenous proteases. Transgenic plants therefore showed no resistance against the bacteria Pseudomonas solanacearum and Pseudomonas syringae, which are highly susceptible to cecropin B in Vitro.

In another approach, substitution analogs of cecropin were produced. In these cecropin analogs, the amino acid sequence was changed but hydrophobic properties and charge density was conserved. One of these substitution analog called Shiva-I with 46 percent homology in amino acid sequence was found to have a more potent lytic activity than another cecropin analog with 95 percent homology. The gene sequence for Shiva-I was introduced into tobacco. Transgenic tobacco seedlings exhibited delayed wilt symptoms, reduced disease severity and reduced mortality after infection with a highly virulent strain of Pseudomonas solanacearum.

Activation of Plant Defence Mechanisms

Phytoalexins are low molecular weight antimicrobial compounds which are synthesized in plants in response to pathogen attack. The synthesis of phytoalexins is in most cases based on complex biosynthetic pathways which are not accessible for genetic engineering yet. An exception are thionins, small cysteine-rich polypeptides which were found in endosperm and leaves of cereals. They were shown to exert antimicrobial activity in vitro. One group obtained high level synthesis of functional thionin in transgenic tobacco leaves transformed with a barley a-thionin gene under control of the CaMV 35S promoter. After inoculation with Pseudomonas syringae pv. tabaci or Pseudomonas syringae pv. syringae, the number of necrotic lesions and the severity of disease symptoms were reduced in leaves of transgenic plants compared to control plants. The level of resistance coincided with the level of thionin expression.

Detoxification of Pathotoxins and Alteration of Target Enzymes

Pathogens that produce phytotoxins causing disease symptoms usually have the capacity to metabolize, i.e. detoxify these compounds. For instance, the dipeptide tabtoxin is thought to be responsible for chlorosis during wildfire disease on tobacco caused by Pseudomonas syringae pv. tabaci. The pathogen protects itself against the toxin by expression of the tabtoxin resistance gene, ttr, which encodes an enzyme that acetylates tabtoxin. Transgenic tobacco plants constitutively expressing the ttr gene under control of the CaMV 35S promoter did not produce the chlorotic halo typical for wildfire disease on non-transgenic plants.

In a comparable strategy, protection against the damaging effects of a bacterial toxin was obtained by transformation of plants with a pathogen-derived enzyme which is unsusceptible to the toxin. Transgenic tobacco plants harboring a gene construct derived from Pseudomonas syringae pv. phaseolicola expressed an ornithine carbomyl transferase which is insensitive to the phaseolotoxin produced by Pseudomonas syringae pv. phaseolicola. Progeny of these plants displayed a reduction in disease symptoms compared to control plants.

Transgenic Baculoviruses

Baculoviruses are pathogens of insects that infect predominantly holometabolous insects. Almost all Lepidopterans, which include many of the world's most serious pests, are susceptible to at least one of the more than 500 baculovirus species. The baculoviruses were first studied for use against forest pests in the 1930s and were used until the advent of chemical pesticides in 1960. The virus particles are generally applied on the foliage by spray. The viruses have no contact action and must therefore be ingested by larvae, thus good coverage of plants is required. Of particular interest for insect pest management is Autographa californica nuclear polyhedrosis virus (AcNPV), originally isolated from the alfalfa looper, and the Bombyx mori nuclear polyhedrosis virus (BmNPV).

The DNA of a baculovirus is enclosed within a so-called nucleocapsid, about 50 by 300 nm in size, which again is enclosed within an occlusion body composed of the protein polyhedrin. These occlusion bodies, each of them containing several nucleocapsids, are referred to as polyhedra. The polyhedra protect the virions in the environment. After ingestion of the polyhedra by an insect, the polyhedrin is dissolved and the nucleocapsids fuse to the midgut cells and migrate to the nucleus, where replication takes place. The produced progeny nucleocapsids spread and infect other cells and tissues, resulting in infection throughout the insect. Nucleocapsids may then be enclosed in new polyhedra synthesized in the nucleus. Before the death of the host, the polyhedra can account for up to 30percent of the insect dry weight. The insect may continue to feed for several weeks before death from viral infection. The cadaver filled with virus ruptures easily, releasing millions of polyhedra onto surrounding foliage and soil.

The fact that baculoviruses take weeks for killing their hosts is one of the most important disadvantages associated with the use of baculoviruses as insecticides. It has restricted their use to crops capable of sustaining some damage without too much economic loss. In order to decrease the killing time of the baculoviruses and the crop damage, genes of different origins have been introduced into AcNPV, including genes encoding insect hormones, insect-specific toxins and insect enzymes.

Field trials with genetically engineered baculoviruses were carried out in England in 1986, 1987 and 1988. The trials examined the persistence of genetically altered AcNPV and were among the first deliberate releases of transgenic organisms. The first recombinant baculovirus with an enhanced insecticidal effect was developed in 1989. The gene of a diuretic hormone which plays an important role in the regulation of the water balance in the insect Manduca sexta (tobacco hornworm) was incorporated into Bombyx mori nuclear polyhedrosis virus (BmNPV). Bombyx mori larvae infected with the transgenic virus died 20percent more quickly than larvae infected with wild-type BmNPV. The expression of a toxin of the North African scorpion Androctonus australis in AcNPV resulted in a 50 percent reduction of feeding damage, and lethal time was reduced by 25 percent compared to wild-type viruses. A toxin from the straw itch mite Pyemotes tritici also introduced into AcNPV to enhance its insecticidal activity reduced time to kill by 30-40percent. The titre of juvenile hormone (JH) in haemolymph determines the course of larval development. A reduction in JH titre controlled by juvenile hormone esterase (JHE) leads to cessation of feeding before a moult and initiates metamorphosis. By expression of modified JHEs in AcNPV, insecticidal viruses with lethal times more than 30 percent lower than with wild-type viruses were generated

Currently, no commercial application of genetically engineered insect viruses is in sight, but their use is thought to have considerable potential. The company American Cyanamid e.g. will conduct field trials with transgenic baculoviruses expressing scorpion toxin in the USA.

The viruses are sometimes wrongly called "Autographa californica' or "Bombyx mori" only. These are the scientific names of the insects from which the viruses were first isolated.

Transgenic Bacteria

Inactivated Pseudomonas Fluorescens for Bt Toxin Delivery

In this approach, a non-pathogenic strain of Pseudomonas fluorescens bacteria is used as host for the Bacillus thuringiensis (Bt) toxin. In an attempt to improve the foliar persistence of Bt insecticidal activity, a delivery system based on recombinant P. fluorescens expressing Bt toxin which are killed by chemicals prior to field release, has been developed by Mycogen Corporation (San Diego, California). When transformed cells are grown in culture, a typical Bt ä-endotoxin crystal is formed within the cell. While still in the fermentation tank, the P. fluorescens cells are chemically treated in order to kill the cells and cause the cell wall to become more rigid through cross-linking of cell wall components. The cell wall of the dead bacteria now serves as a protective microcapsule for the enclosed ä-endotoxin. This Bt toxin delivery system has been called CellCap by Mycogen.

Since no living transgenic microorganisms are released into the environment, this approach has the advantage of relative freedom from environmental and safety concerns associated with field releases of living transgenic organisms. The encapsulated Mycogen product with a lepidopteran active ä-endotoxin was the first recombinant product approved for outdoor testing.

Field experiments conducted during 1988 suggested that a foliar application of this engineered microbe protected cabbage from lepidopteran pests for 7 days, whereas insecticidal activity from traditional Bt sprays dissipated by 3 days post-application. This resulted in better insect control, and therefore higher yields. Similar tests have been conducted with engineered microbes targeted against the Colorado potato beetle on potatoes. It was shown that the protected ä-endotoxin could be applied at two thirds the toxin rate and still maintain higher levels of insect control than the full rate delivered by Bacillus thuringiensis.

Bacterial Endophytes Expressing Bacillus Thuringiensis Toxins

The term endophyte is used for plant-associated microorganisms that live within their host plants. Genetic engineering of nonpathogenic endophytes presents an opportunity for delivery of biopesticides in the plant, ie. the production of the pesticidal agent within host plant tissues without direct genetic manipulation of the host plant. Most approaches make use of Bacillus thuringiensis (Bt) toxin. The major drawback in using Bacillus thuringiensis directly as a biopesticide lies in the fast degradation of the Bt toxin after spraying the bacteria with the Bt toxin crystal outside the spore. A systemic delivery system using bacteria living within the plant would protect the Bt toxin from degradation, providing consistent levels of protection throughout the growing season by the presence of the active ingredient produced by the endophyte. Areas difficult to reach with chemical sprays (e.g., the inside of stalks) would be better protected from plant-feeding insects by endophytes expressing Bt toxin.

A number of plant-associated bacteria have been transformed with Bt toxin genes, including Pseudomonas fluorescens, Pseudomonas cepacia, Rhizobium meliloti and Rhizobium leguminosarum. In the following, the use of transgenic Clavibacter as biopesticide will be described in more detail, as this application is one of the most advanced in this field.

The Gram-negative bacterium Clavibacter xyli ssp. cynodontis (Cxc) is a common endophyte living in the xylem of Bermudagrass (Cynodon dactylon) and has been found in this host in the United States, Japan, Taiwan, and France. It has also been found in a few weed species, and it can colonize maize, rice, sorghum, oats, white millet and sudangrass through artificial inoculation. The gene coding for the CrylA(c) protein from Bt ssp kurstaki was inserted into the chromosome of a wild-type Cxc. Unlike Bt, transgenic Cxc do not release the Bt toxin. Instead, the cell must be digested by the insect in order to release the active ingredient in the gut.

The observed effects are similar to those observed in larvae ingesting Bt. Field studies have demonstrated that a wide range of commercial field and sweet maize hybrids can be inoculated successfully, with 80-100 percent of the inoculated seeds producing vigorous, endophyte colonized plants. The European corn borer (ECB), Ostrinia nubilalis, is cosmopolitan in its distribution and a significant pest in maize.

Maize plants inoculated with recombinant Cxc; and artificially infested with ECB larvae sustained up to 80 percent less damage than plants inoculated with either wild type Cxc or uninoculated plants. C= was the first recombinant microorganism expressing Bt toxin which has been approved for field testing.

INTRODUCTION: CROP PRODUCTION AND CROP LOSSES

The present human population of the world is 5.7 billion, increasing at an annual rate of 1.6 percent. According to the latest estimates, the rate of population growth will decrease slowly during the next century, and the world population is expected to stabilize at about 11.5 billion shortly after the year 2100.

The rapid population growth in general and the fact that most of the population growth takes place in developing countries rises the question how these additional people can be fed. Basically, there are two different possibilities to increase food production: Enlarging crop yields on already cultured land, or expanding the area of cultured land on previously uncultured land.

Reviewing the development of crop production since 1965 reveals a massive increase. The differences in production between 1965 and 1988-90 (average of three year period) of the eight most important food and cash crops. The production of rice, wheat and maize, which are the most important crops for human nutrition, doubled during the last 25 years. Figure indicates that most of the production increase was due to higher yields per area. Higher yields per area were possible because of the introduction of high-yielding varieties, improved cultivation techniques, fertilizer use and chemical crop protection measures - a development which has been termed "green revolution".

The intensification of crop production improved the level of food consumption in most developing countries during the last 30 years despite a world population growth of 2.4 billion in the same period. However, the intensification also led to a number of disadvantageous side-effects like erosion, deterioration of soil quality, salinization, and others. Every year, about 1 percent of the cultured area has to be replaced because of these reasons. Especially in developed countries, the use of agrochemicals is a growing public concern. The amount of crop protection products used has risen steadily during the past 30 years. Presently, more than 25 billion US$ per year are spent for herbicides, fungicides, insecticides, soil disinfectants and growth regulators. In Western Europe, the most crop protection products are used per area of arable land, followed by the USA. The expenditure per area in these and the other principal farming regions of the world.

As can be seen in figure, losses prevented by chemical crop protection are about 27 percent (weeds 16 percent, animal pests 7 percent, diseases 4 percent) of attainable production in the eight principal food and cash crops. Nevertheless, about 42 percent of total production in these crops are lost to weeds (13 percent), animal pests (16 percent) and pathogens (13 percent), representing a market value of about 244 billion US$.

Despite the use of pesticides, not only the absolute amount of losses, but also relative losses have increased by about 3 to 10 percent in the past 30 years. This is thought to be due to the higher susceptibility of the high-yielding varieties, the development of resistance to chemical pesticides in pest and pathogen populations, the use of fertilizers which not only promote growth but also increase susceptibility, and the extension of production into regions where the pressure from pests, pathogens and weeds is higher. The losses have increased predominantly in developing countries, where appropriate crop protection measures often cannot be taken because of their high price and/or lack of knowledge.

Plant genetic engineering may contribute substantially to the development of new pest and disease resistant varieties by increasing the speed of conventional breeding and by transferring useful genes from various plants and other organisms into crops. The principal methods used to achieve these goals are described. The present state of the art of genetic engineering for pest and disease resistance and applications in development or already realized will be presented. A brief discussion of safety issues raised and the regulatory framework for field releases and market introduction complements the topic. As not only plants, but also other organisms like bacteria or viruses are genetically engineered in order to prevent crop losses.

Chapter 4

Plant Cells

Most cells are not visible with the naked eye. However, with microscopes of various types, plant cells can be readily viewed and studied. In young parts of plant and fruits, cell shapes are generally round, while in older sections, the cells are somewhat boxlike with up to 14 sides as they become packed together.

A plant cell is bounded by a cell wall and the living portion of the cell is within the walls and is divided into two portions: the nucleus, or central control center; and the cytoplasm, a fluid in which membrane bound organelles are found. Between the primary cell walls of adjacent plant cells, lies a pectic middle lamella. There can be a secondary cell wall which would be located just to the inside of the primary wall. Both walls consist mainly of cellulose, but the secondary cell wall may contain lignin and other substances. The outer boundary of the protoplasm (cytoplasm and nucleus) is a sandwich-like, flexible plasma membrane. This membrane regulates what enters and leaves the plant cell. Plant cell organelles include: endoplasmic reticulum, with and without ribosomes attached; Golgi bodies, mitochondria, and plastids. Plastids are chloroplasts, chromoplasts or leucoplasts—depending on the colour and likewise the function. Chloroplasts are of specific interest to those studying plants. A plant cell also, obviously, contains a nucleus which is bounded by a nuclear envelope with pores. The pores in the nuclear envelope allow for movement of substances in and out of the nucleus. Within the nucleus is a number of chromosomes. The number present is specific to the organism and it will be later noted how sex cells contain one-half the number of chromosomes, and restore chromosome number upon fertilization. All of these organelles and the nucleus are suspended in the cytoplasm. The cytoplasm has movements that are referred to as cytoplasmic streaming or cyclosis. The particular function of the other organelles contained in plant cells can be reviewed below:

1. The nucleus is in the center of most cells. Some cells contain multiple nuclei, such as skeletal muscle, while some do not have any, such as red blood cells. The nucleus is the largest membrane-bound organelle. Specifically, it is responsible for storing and transmitting genetic information. The nucleus is surrounded by a selective nuclear envelope. The nuclear envelope is composed of two membranes joined at regular intervals to form circular openings called nuclear pores. The pores allow RNA molecules and proteins modulating DNA expression to move through the pores and into the cytosol. The selection process is controlled by an energy-dependent process that alters the diameter of the pores in response to signals. Inside the nucleus, DNA and proteins associate to form a network of threads called chromatin. The chromatin becomes vital at the time of cell division as it becomes tightly condensed thus forming the rodlike chromosomes with the enmeshed DNA. Inside the nucleus is a filamentous region called the nucleolus. This serves as a site where the RNA and protein components of ribosomes are assembled. The nucleolus is not membrane bound, but rather just a region.
2. Ribosomes are the sites where protein molecules are synthesized from amino acids. They are composed of proteins and RNA. Some ribosomes are found

bound to granular endoplasmic reticulum, while others are free in the cytoplasm. The proteins synthesized on ribosomes bound to granular endoplasmic reticulum are transferred from the lumen (open space inside endoplasmic reticulum) to the golgi apparatus for secretion outside the cell or distribution to other organelles. The proteins that are synthesized of free ribosomes are released into the cytosol.

3. The endoplasmic reticulum (ER) is collectively a network of membranes enclosing a singular continuous space. As mentioned earlier, granular endoplasmic reticulum is associated with ribosomes (giving the exterior surface a rough, or granular appearance). Sometimes granular endoplasmic reticulum is referred to as rough ER. The granular ER is involved in packaging proteins for the golgi apparatus. The agranular, or smooth, ER lacks ribosomes and is the site of lipid synthesis. In addition, the agranular ER stores and releases calcium ions (Ca^{2+}).
4. The golgi apparatus is a membranous sac that serves to modify and sort proteins into secretory/transport vesicles. The vesicles are then delivered to other cell organelles and the plasma membrane. Most cells have at least one golgi apparatus, although some may have multiple. The apparatus is usually located near the nucleus.
5. Endosomes are membrane-bound tubular and vesicular structures located between the plasma membrane and the golgi apparatus. They serve to sort and direct vesicular traffic by pinching off vesicles or fusing with them.
6. Mitochondria are some of the most important structures in the cell. They are they site of various chemical processes involved in the synthesis of energy packets called ATP (adenosine triphosphate).

Each mitochondrion is surrounded by two membranes. The outer membrane is smooth, while the inner one is folded into tubule structures called cristae. Mitochondria are unique in that they contain small amounts of DNA containing the genes for the synthesis of some mitochondrial proteins. The DNA is inherited solely from the mother. Cells with greater activity have more mitochondria, while those that are less active have less need for energy producing mitochondria.

7. Lysosomes are bound by a single membrane and contain highly acidic fluid. The fluid acts as digesting enzymes for breaking down bacteria and cell debris. They play an important from in the cells of the immune system.
8. Peroxisomes are also bound by a single membrane. They consume oxygen and work to drive reactions that remove hydrogen from various molecules in the form of hydrogen peroxide. They are important in maintaining the chemical balances within the cell.
9. The cytoskeleton is a filamentous network of proteins that are associated with the processes that maintain and change cell shape and produce cell movements in animal and bacteria cells. In plants, it is responsible for maintaining structures within the plant cell, rather then whole cell movement. The cytoskeleton also forms tracks along which cell organelles move propelled by contractile proteins attached to their various surfaces. Like a little highway infrastructure inside the cell. Three types of filaments make up the cytoskeleton.
   a. Microfilaments are the thinnest and most abundant of the cytoskeleton proteins. They are composed of actin, a contractile protein, and can be assembled and disassembled quickly according to the needs of the cell or organelle structure.

b. Intermediate filaments are slightly larger in diameter and are found most extensively in regions of cells that are going to be subjected to stress. Once these filaments are assembled they are not capable of rapid disassembly.

c. Microtubules are hollow tubes composed of a protein called tubulin. They are the thickest and most rigid of the filaments. Microtubules are present in the axons and long dendrite projections of nerve cells. They are capable of rapid assembly and disassembly according to need. Microtubules are structured around a cell region called the centrosome, which surrounds two centrioles composed of 9 sets of fused microtubules. These are important in cell division when the centrosome generates the microtubluar spindle fibers necessary for chromosome separation.

10. *Chloroplasts:* It is necessary to note a bit about the form of chloroplasts. Inside a chloroplast is a matrix called the stroma. Enzymes are found in the stroma as well as grana—stacks of coin-shaped discs, called thylakoids. It is within the thylakoids that photosynthesis takes place. Note that chloroplasts, like mitochondria contain their own DNA. They do rely on proteins from the nucleus, and are considered semi-autonomous organelles.
11. Vacuoles: Plant cells are also notorious for having huge vacuoles. Up to 90percent of the volume of a mature cell may be taken up by a single large vacuole or several vacuoles. The vacuole is bound by a special membrane, called the tonoplast, and contains cell sap—which is composed of dissolved substances and may include pigments.

CELL CYCLE

The cell cycle contains the process in which cells are either dividing or in between divisions. Cells that are not actively dividing are said to be in interphase, which has three distinct periods of intense activity that precedes the division of the nucleus, or mitosis. The division of the rest of the cell occurs as an end result of mitosis and this process occurs in regions of active cell division, called meristems. Meristems.

Mitosis is a process within the cell cycle that is divided into four phases which we will sum up here:

1. Prophase—the chromosomes and their usual two-stranded nature becomes apparent, the nuclear envelope breaks down.
2. Metaphase—the chromosomes become aligned at the equator of the cell. A spindle composed of spindle fibers is developed and some attach to the chromosomes at their centromere.
3. Anaphase—the sister chromatids of each chromosome, that is now called the daughter chromosomes, separate lengthwise and each group of daughter chromosomes migrates to the opposite ends of the cell.
4. Telophase—the groups of daughter chromosomes are grouped within a developing nuclear envelope which makes them separate nuclei. A wall forms between the two sets of daughter chromosomes thus creating two daughter cells.

In plants, as the cell wall is developing, droplets or vesicles of pectin merge forming a cell plate that eventually will become the middle lamella of the new cell wall.

PLANT CELLS COMPARED WITH ANIMAL CELLS

Animal cells do not have a cell wall. Instead of a cell wall, the plasma membrane (usually called cell membrane when discussing animal cells) is the outer boundary of animal cells. Animal tissues therefore require either external or internal support from some kind of skeleton. Frameworks of rigid cellulose fibrils thicken and strengthen the cell walls of higher plants. Plasmodesmata that connect the protoplasts of higher plant cells do not have a counterpart in the animal cell model. During telophase of mitosis, a cell plate is formed as the plant cell begins its division. In animal cells, the cell pinches in the center to form two cells; no cell plate is laid down. Centrioles are generally not found in higher plant cells, while they are found in animal cells. Animal cells do not have plastids, which are common in plant cells (chloroplasts). Both cell types have vacuoles, however, in animal cells vacuoles are very tiny or absent, while in plant cells vacuoles are generally quite large.

PLANT TISSUES

Plants are composed of three major organ groups: roots, stems and leaves. As we know from other areas of biology, these organs are comprised of tissues working together for a common goal (function). In turn, tissues are made of a number of cells which are made of elements and atoms on the most fundamental level. In this section, we will look at the various types of plant tissue and their place and purpose within a plant. It is important to realize that there may be slight variations and modifications to the basic tissue types in special plants.

Plant tissues are characterized and classified according to their structure and function. The organs that they form will be organized into patterns within a plant which will aid in further classifying the plant. A good example of this is the three basic tissue patterns found in roots and stems which serve to delineate between woody dicot, herbaceous dicot and monocot plants.

Meristematic Tissues

Tissues where cells are constantly dividing are called meristems or meristematic tissues. These regions produce new cells. These new cells are generally small, six-sided boxlike structures with a number of tiny vacuoles and a large nucleus, by comparison. Sometimes there are no vacuoles at all. As the cells mature the vacuoles will grow to many different shapes and sizes, depending on the needs of the cell. It is possible that the vacuole may fill 95percent or more of the cell's total volume.

There are three types of meristems:

1. Apical Meristems
2. Lateral Meristems
3. Intercalary Meristems

Apical meristems are located at or near the tips of roots and shoots. As new cells form in the meristems, the roots and shoots will increase in length. This vertical growth is also known as primary growth. A good example would be the growth of a tree in height. Each apical meristem will produce embryo leaves and buds as well as three types of primary meristems: protoderm, ground meristems, and procambium. These primary meristems will produce the cells that will form the primary tissues.

Lateral meristems account for secondary growth in plants. Secondary growth is generally horizontal growth. A good example would be the growth of a tree trunk in girth. There are two types of lateral meristems to be aware of in the study of plants.

The vascular cambium, the first type of lateral meristem, is sometimes just called the cambium. The cambium is a thin, branching cylinder that, except for the tips where the apical meristems are located, runs the length of the roots and stems of most perennial plants and many herbaceous annuals. The cambium is responsible for the production of cells and tissues that increase the thickness, or girth, of the plant.

The cork cambium, the second type of lateral meristem, is much like the vascular cambium in that it is also a thin cylinder that runs the length of roots and stems. The difference is that it is only found in woody plants, as it will produce the outer bark.

Both the vascular cambium and the cork cambium, if present, will begin to produce cells and tissues only after the primary tissues produced by the apical meristems have begun to mature.

Intercalary meristems are found in grasses and related plants that do not have a vascular cambium or a cork cambium, as they do not increase in girth. These plants do have apical meristems and in areas of leaf attachment, called nodes, they have the third type of meristematic tissue. This meristem will also actively produce new cells and is responsibly for increases in length. The intercalary meristem is responsible for the regrowth of cut grass.

There are other tissues in plants that do not actively produce new cells. These tissues are called nonmeristematic tissues. Nonmeristematic tissues are made of cells that are produced by the meristems and are formed to various shapes and sizes depending on their intended function in the plant. Sometimes the tissues are composed of the same type of cells throughout, or sometimes they are mixed. There are simple tissues and complex tissues to consider, but we will start with the simple tissues for the sake of discussion.

Simple Tissues

There are threé basic types, named for the type of cell that makes up their composition.

1. Parenchyma cells form parenchyma tissue. Parenchyma cells are the most abundant of cell types and are found in almost all major parts of higher plants. These cells are basically sphere shaped when they are first made. However, these cells have thin walls, which flatten at the points of contact when many cells are packed together. Generally, they have many sides with the majority having 14 sides. These cells have large vacuoles and may contain various secretions including starch, oils, tannins, and crystals. Some parenchyma cells have many chloroplasts and form the tissues found in leaves. This type of tissue is called chlorenchyma. The chief function of this type of tissue is photosynthesis, while parenchyma tissues without chloroplasts are generally used for food or water storage. Additionally, some groups of cells are loosely packed together with connected air spaces,

such as in water lilies, this tissue is called aerenchyma tissue. These type of cells can also develop irregular extensions of the inner wall which increases overall surface area of the plasma membrane and facilitates transferring of dissolved substances between adjacent cells. Parenchyma cells can divide if they are mature, and this is vital in repairing damage to plant tissues. Parenchyma cells and tissues comprise most of the edible portions of fruit.

2. Collenchyma cells form collenchyma tissue. These cells have a living protoplasm, like parenchyma cells, and may also stay alive for a long period of time. Their main distinguishing difference from parenchyma cells is the increased thickness of their walls. In cross section, the walls looks uneven. Collenchyma cells are found just beneath the epidermis and generally they are elongated and their walls are pliable in addition to being strong. As a plant grows these cells and the tissues they form, provide flexible support for organs such as leaves and flower parts. Good examples of collenchyma plant cells are the 'strings' from celery that get stuck in our teeth.

3. Sclerenchyma cells form sclerenchyma tissue. These cells have thick, tough secondary walls that are imbedded with lignin. At maturity, most sclerenchyma cells are dead and function in structure and support. Sclerenchyma cells can occur in two forms:

    a. Sclereids are sclerenchyma cells that are randomly distributed throughout other tissues. Sometimes they are grouped within other tissues in specific zones or regions. They are generally as long as they are wide. An example, would be the gritty texture in some types of pears. The grittiness is due to groups of sclereid cells. Sclereids are sometimes called stone cells.

b. Fibers are sometimes found in association with a wide variety of tissues in roots, stems, leaves and fruits. Usually fiber cells are much longer than they are wide and have a very tiny cavity in the center of the cell. Currently, fibers from over 40 different plant families are used in the manufacture of textiles, ropes, string and canvas goods to name a few.

Secretory Cells and Tissues

As a result of cellular processes, substances that are left to accumulate within the cell can sometimes damage the protoplasm. Thus it is essential that these materials are either isolated from the protoplasm in which they originate, or be moved outside the plant body. Although most of these substances are waste products, some substances are vital to normal plant functions. Examples: oils in citrus, pine resin, latex, opium, nectar, perfumes and plant hormones. Generally, secretory cells are derived from parenchyma cells and may function on their own or as a tissue. They sometimes have great commercial value.

Complex Tissues

Tissues composed of more than one cell type are generically referred to as complex tissues. Xylem and phloem are the two most important complex tissues in a plant, as their primary functions include the transport of water, ions and soluble food substances throughout the plant. While some complex tissues are produced by apical meristems, most in woody plants are produced by the vascular cambium and is often referenced as vascular tissue. Other complex tissues include the epidermis and the periderm. The epidermis consists primarily of parenchyma-like cells and forms a protective covering for all plant organs. The epidermis includes specialized cells that allow for the movement of water and gases in and out of the plant, secretory glands, various hairs, cells in which crystals are accumulated and isolated, and other cells that increase absorption in the roots. The periderm is mostly cork cells and therefore forms the outer bark of woody plants. It is considered to be a complex tissue because of the pockets of parenchyma cells scattered throughout.

Xylem

Xylem is an important plant tissue as it is part of the 'plumbing' of a plant. Think of bundles of pipes running along the main axis of stems and roots. It carries water and dissolved substances throughout and consists of a combination of parenchyma cells, fibers, vessels, tracheids and ray cells. Long tubes made up of individual cells are the vessels, while vessel members are open at each end. Internally, there may be bars of wall material extending across the open space. These cells are joined end to end to form long tubes. Vessel members and tracheids are dead at maturity. Tracheids have thick secondary cell walls and are tapered at the ends. They do not have end openings such as the vessels. The tracheids ends overlap with each other, with pairs of pits present. The pit pairs allow water to pass from cell to cell. While most conduction in the xylem is up and down, there is some side-to-side or lateral conduction via rays. Rays are horizontal rows of long-living parenchyma cells that arise out of the vascular cambium. In trees, and other woody plants, ray will radiate out from the center of stems and roots and in cross-section will look like the spokes of a wheel.

Phloem

Phloem is an equally important plant tissue as it also is part of the 'plumbing' of a plant. Primarily, phloem carries dissolved food substances throughout the plant. This conduction system is composed of sieve-tube member and companion cells, that are without secondary walls. The parent cells of the vascular cambium produce both xylem and phloem. This usually also includes fibers, parenchyma and ray cells. Sieve tubes are formed from sieve-tube members laid end to end. The end walls, unlike vessel members in xylem, do not have openings. The end walls, however, are full of small pores where cytoplasm extends from cell to cell. These porous connections are called sieve plates. In spite of the fact that their cytoplasm is actively involved in the conduction of food materials, sieve-tube members do not have nuclei at maturity. It is the companion cells that are nestled between sieve-tube members that function in some manner bringing about the conduction of food. Sieve-tube members that are alive contain a polymer called callose. Callose stays in solution as long at the cell contents are under pressure. As a repair mechanism, if an insect injures a cell and the pressure drops, the callose will precipitate. However, the callose and a phloem protein will be moved through the nearest sieve plate where they will for a plug. This prevents further leakage of sieve tube contents and the injury is not necessarily fatal to overall plant turgor pressure.

Epidermis

The epidermis is also a complex plant tissue, and an interesting one at that. Officially, the epidermis is the outermost layer of cells on all plant organs (roots, stems, leaves). The epidermis is in direct contact with the environment and therefore is subject to environmental conditions and constraints. Generally, the epidermis is one cell layer thick, however there are exceptions such as tropical plants where the layer may be several cells thick and thus acts as a sponge. Cutin, a fatty substance secreted by most epidermal cells, forms a waxy protective layer called the cuticle.

The thickness of the cuticle is one of the main determiners of how much water is lost by evapouration. Additionally, at no extra charge, the cuticle provides some resistance to bacteria and other disease organisms. Some plants, such as the wax palm, produce enough cuticle to have commercial value: carnauba wax. Other wax products are used as polishes, candles and even phonographic records. Epidermal cells are important for increasing absorptive surface area in root hairs. Root hairs are essentially tubular extensions of the main root body composed entirely of epidermal cells. Leaves are not left out. They have many small pores called stomata that are surrounded by pairs of specialized epidermal cells called guard cells. Guard cells are unique epidermal cells because they are of a different shape and contain chloroplasts. There are other modified epidermal cells that may be glands or hairs that repel insects or reduce water loss.

Periderm

In woody plants, when the cork cambium begins to produce new tissues to increase the girth of the stem or root the epidermis is sloughed off and replaced by a periderm. The periderm is made of semi-rectangular and boxlike cork cells. This will be the outermost layer of bark. These cells are dead at maturity. However, before the cells die, the protoplasm secretes a fatty substance called suberin into the cell walls. Suberin makes the cork cells waterproof and aids in protecting tissues beneath the bark. There are parts of the cork cambium that produce pockets of loosely packed cork cells. These cork cells do not have suberin imbedded in their cell walls. These loose areas are extended through the surface of the periderm and are called lenticels. Lenticels function in gas exchange between the air and the stem interior. At the bottom of the deep fissures in tree bark are the lenticels.

PLANT STEMS

A woody twig, or stem, is an axis with leaves attached. The leaves are arranged in various ways around and on the axis. You may hear them described as alternate, or alternately arranged, opposite or oppositely arranged, or if they are found in groups of three or more they may be referred to as whorled. The region, just a general area in this case, where the leaves attach to the stem are called nodes. The region of stem between two nodes is called the internode. The leaf blade is attached to the stem via a stalk called the petiole. In the angle, or axil, formed between the petiole and the stem you will find the axillary bud. Axillary refers to a structure that forms an armpit, just for trivia's sake. These buds can become new branches or they may have tissues that will form into flowers for the next season. Most buds are protected by bud scales which fall off as bud tissue begins to grow. In general, at the tip of a twig a terminal (or ending) bud is present. It is larger than the axillary buds and produces tissues to extend twig length during the growing season. When the bud scales of a terminal bud fall off they leave scars on the twig. You can calculate the age of a twig by counting up the terminal bud scale scars. There are other scars on twigs that may look like terminal bud scars that are left by paired appendages called stipules which are found at the base of a petiole in the axil.

Trees and shrubs that lose their leaves every year, deciduous plants, have characteristic leaf scars with dormant, or not active, axillary buds directly above them. Sometimes tiny bundle scars can be seen. These scars are found in the leaf scar and mark the location of food and water conducting tissues. The shape and arrangement of the bundle scars can help distinguish deciduous trees in the winter months when the leaf structures are absent.

THE ORIGIN AND DEVELOPMENT OF STEMS

Recall that the apical meristem is responsible for vertical growth, or increase in length of a stem. Prior to the start of the growing season, the cells in the apical meristem are dormant. The apical meristem is protected at the tip of the twig, by the covering bud scales and by the leaf primordial. The leaf primordia are tiny embryonic leaves that will develop into mature leaves after bud scales drop off and growth commences. When a seed germinates or a bud begins to grow, the cells in the apical meristem undergo mitosis. From these cells three primary meristems will develop:

1. The outermost meristem is the protoderm, which gives rise to the epidermis. This layer is usually one cell thick and becomes coated with a waxy cuticle.
2. The second layer is the procambium, which is a cylinder of strands. This layer gives rise to the primary xylem and primary phloem cells.

3. The innermost meristem is the ground meristem from which arises two tissues composed of parenchyma cells. The tissue in the center of the stem is the pith. These cells are large and may break down shortly after being formed which leaves a cylindrical hollow area. If they do not break down, they will be compressed by new additions to the plant girth by other meristems. The second parenchyma tissue that arises is called cortex. Cortex may be quite extensive and also crushed or replaced in woody stems. The function of both tissues is food storage. If chloroplasts are present the tissues may function in producing food.

It is important to note that all five of the above mentioned tissues—epidermis, primary xylem, primary phloem, pith and cortex—are produced by the apical meristem and are thus primary tissues as the plant is increasing in length. Xylem and phloem tissue branch off from the main vascular cylinder and enter into the leaf or bud. Each branching of vascular tissue is called a trace. Each trace branch leaves a small thumbnail shaped gap in the cylinder of tissue and are called leaf gaps and bud gaps.

In between the primary xylem and primary phloem a thin band of cells retains its meristematic nature. This band becomes the vascular cambium of one of the two lateral meristems.

In woody plants, and some others, a second cambium arises from the cortex or sometimes the epidermis or phloem. The second cambium is called the cork cambium or phellogen and is responsible for producing cork cells. Recall that the cork cells become filled with suberin which waterproofs the cells. The resulting cork tissue constitutes the out bark of woody plants and functions to reduce water loss and to protect the stem against mechanical injury. We will revisit the role of cork later on in discussing biotechnology and propagation. For now, though, understand that cork tissue cuts off food and water supplies to the epidermis which results in a sloughing off. Also understand that cork tissues do not form a solid cylinder around the exterior of a woody stem. This is to allow vital gas exchange with the environment.

Before we go on, it is important to remember the difference between monocots and dicots, the two main divisions of flowering plants. Most of the distinguishing revolves around the seed leaves, which are called cotyledons. Cotyledons function in storing food needed by the young seedling until true leaves grow and are able to take over the food supplying function.

1. Monocotyledon (monocot) plants—These plants form from seeds that have one embryonic seed leaf (hence the 'mono' in monocot).
2. Dicotyledon (dicot) plants—These plants form from seeds that have two (hence the 'di' in dicot) embryonic seed leaf.

Cone bearing trees, conifers such as pines, have multiple cotyledons, usually eight, in their seed structure.

There are four tissue patterns to be aware of in the study of plants.

1. Steles—steles are a central cylinder in most younger stems and roots, composed of primary xylem, phloem and the pith, if present. Sometimes referred to as eusteles, which are vascular bundles in higher vascular plants.
2. Herbaceous Dicotyledonous Stems—Herbaceous refers to non-woody plants. Plants that die after going from seed to maturity are called annuals. In general, most monocots are annuals, but there are annual dicot plants as well. Annual dicots are mostly composed of primary tissues, although there may be some minimal secondary growth. Remember, the plant only lives a year, so extensive secondary growth, or increases in width, really doesn't make

sense as far as using the plant resources. A cross-section of a herbaceous dicot stem will show discrete patches of xylem and phloem, vascular bundles, that are arranged in a proper ring separating the cortex and the pith. If secondary xylem and phloem are to emerge, they will arise from between the two primary tissues. Monocots will be discussed shortly.

3. Woody Dicotyledonous Stems—Wood is essentially secondary xylem growth. These stems look similar to herbaceous dicot stems up until the vascular cambium and the cork cambium start functioning. The differences are then quite obvious. While some tropical trees demonstrate year round secondary growth, most trees in temperate climates grow in the spring and summer and cease through the winter. In the springtime, when water and resources are plentiful, the vascular cambium produces large xylem cells. During the summer months when resources and water may be lacking or reduced, the xylem cells are small. Pressed up against the large, light coloured xylem cells, the small xylem cells look like a thin dark ring.

   One year of xylem growth, called an annual ring, can be measured as the distance between the dark rings—or the distance between summer xylem growth. Summer growth is called summer wood while the large spring cells are called spring wood. Much can be learned about the local environmental conditions through the years by looking at tree rings. If water is plentiful the rings will be wider than usual. Years with fires and blights will be evident, as well as insect infestations and fungal infections. All this by looking at a cross section of a tree. In conifers, vessels and fibers are absent and thus the wood consists mainly of tracheids. It is important therefore to remember that environmental conditions affect xylem production and the dark rings may not be completely visible, one year's growth is what constitute an annual ring, not just dark circles.

The vascular cambium produces more xylem than phloem. In fact, the phloem will be difficult to locate as the cells are thinner than xylem and more likely to collapse under the pressure of the cambiums. Phloem grows to the outside of the vascular cambium and xylem grows to the inside. The oldest xylem is in the very center of the stem/trunk. The wood in the center is called heartwood. It is usually darker as the vessels and tracheids are filled with old resins, gums and tannins. The younger wood where the xylem is still functioning is toward the outside of the stem nearest the cambium and is lighter in colour. This younger wood is called sapwood. The main role of heartwood is structure and support, since it is unable to conduct water and nutrients. The heartwood sometimes rots out of a otherwise living tree. Sapwood develops at roughly the same rate that heartwood is 'retiring' and thus vital conducting functions are not compromised. Recall that conifers do not have vessels or fibers and are primarily tracheids. Conifers have resin canals scattered throughout the xylem tissue. Conifers are primarily considered to be softwoods while the wood of woody dicot trees are considered to be hardwood.

Bark is all of the tissues outside of the cambium, including the phloem. Some have gone so far as to distinguish between inner bark—primary and secondary phlolem and outer bark—the periderm, which consists of cork tissue and cork cambium. The cells in these layers only function briefly as they usually become crushed and then slough off. New layers are annually produced by the cambiums. The youngest phloem cells are the ones nearest the vascular cambium and are most active in transporting nutrients, sugars and water. Mature bark may be composed of alternating layers of crushed phloem and cork.

Monocotyledonous Stems: These plants are usually grass or grass-like and do not grow to great size. Monocot stems do not have vascular cambiums or cork cambiums, as growth will not be laterally. The vascular bundles produced by the procambium are scattered throughout the stem, rather than organized in rings as in woody dicot stems. Every bundle is oriented with the xylem toward the center of the stem and the phloem toward the stem surface. The xylem in the vascular bundle generally consists of two large vessels with some small vessels in between them, while the phloem consists of sieve tubes and companion cells. The entire vascular bundle is wrapped in a sheath of sclerenchyma cells. The background tissue between vascular bundles is not divided into cortex and pith in monocots, but they do have similar function and appearance as the parenchyma cells in cortex and pith. The concentration of bundles and bands of sclerenchyma cells, give the stem the flexibility and strength to withstand the elements—such as a summer rainstorm. In grasses, there is an intercalary meristem at the base o each internode which contributes to growth in length, like apical meristems. During the growing season, the stems of the grasses elongate rapidly. Because there is no vascular cambium that would produce tissues to increase the girth of the plant, the growth is columnar with very little variation in diameter between the top and the bottom of the plant.

Palm trees are special, because they grow to considerable size, however this is primarily due to the subsequent division and growth of their parenchyma cells. All this growth occurs without a true cambium developing. Other monocot stems have adaptations that allow for specialized growth. Monocot fibers, such as manila hemp and sisal, come from stems and leaves and are used for commercial products however, their fibers are not as strong as dicot fibers.

Rhizomes—horizontal stems that grow beneath the ground, but near the surface of the soil. They resemble roots, but are actually modified stems with scale-like leaves and buds at each axillary node. In addition, adventitious roots are produced along the rhizome on the lower surface in order to increase absorption surface area.

Runners and Stolons—Runners are horizontal stems that grow above ground, usually along the surface (compare with rhizomes). Strawberry plants produce runners after the first flowering of the season, they may extend out up to 3 feet or more beyond the parent plant. Along the runner, adventitious buds will develop in order to propagate new plants. Stolons are similar to runners, except that they grow roughly vertically beneath the surface of the soil. Irish potato plants have tubers at the tips of stolons.

Tubers—Tubers develop at the tips of stolons. The plant accumulates food at the stolon and the area swells at the internodes. When the tuber is mature the stolon will die and the 'eyes' of the potato are actually nodes arranged in a spiral around the modified stem. Each eye has an axillary bud in the axil of a tiny leaf, which is not always visible in maturing tubers.

Bulbs—These are actually large buds with a small stem at the lower end that is surrounded with fleshy leaves. Onions, irises and tulips are good examples of bulbs and their main function is food storage.

Corms—On first glance you might think these guys are bulbs, however, the differences lie beneath the thin layer of leaves covering the outside of the corm. Adventitious roots form beneath the fullness of the base. Corms function in storing food. Crocuses and gladioli are good examples of plants with corms.

Cladophylls—These are usually called the prickly part of a cactus. Cladophylls are flattened and somewhat leaf-like in appearance. They center of each cladophyll usually has a node with small scalelike leaves complete with axillary buds. The scaly look to asparagus are cladophylls. These specialized stems are not only restricted to cacti, but are found in some orchids and greenbriars.

Other Specialized Stems—Cacti usually have modifications in their stem or 'trunk' structure in order to hold extra water. Other stems may be modified into thorns or briars. It is important to remember that not all thorn-like structures are stems! Raspberry and rose prickles are extensions of their epidermis and are neither thorns nor spines. Other stems are modified for climbing, such as tendrils and ramblers.

Stems are vital to the human cause. They provide building materials, paper products, food and much, much more! Stop and think of how many things you encounter in a day that is either made of wood or plant fiber or a derivative product, chances are good the are a stem derivative.

ROOTS

Upon seed germination, the embryo root, called the radicle, grows and develops into the first root. The radicle may thicken into a taproot with many branching roots, or it may develop into many adventitious roots. The direct opposite of a taproot system is a fibrous root system. This develops out of the many adventitious roots. In diameter, the roots in a fibrous system are very fine. There are many mature plants that have a combination system, which means there is a main taproot with many branching fibrous roots attached. Root hairs, or extensions of the epidermis, significantly increase the contact surface area of the root system. This allows for more exchange with the surrounding soil.

In general, most dicot plants (peas, carrots), or two seed-leaf plants, have taproot systems while monocot plants (corn, grasses), or one seed-leaf plants, have fibrous root systems. Additional differences between dicots and monocots will be discussed later on.

ROOT STRUCTURE

Historically, developing roots have been categorized into four zones of development. These are not strict zones, but rather regions of cells that gradually develop into those of the next region. The zones vary widely as far as extent and levels of development.

Regions of root development:

1. Root cap
2. Region of cell division

3. Region of elongation
4. Region of maturation

We will discuss each region in greater detail.

ROOT CAP

In some plants the root cap is quite large and obvious, while in others it is nearly impossible to find. The root cap is made of parenchyma cells that form a thimble shape, as a covering for the tip of each root. The cap serves several functions. The main function being protection as the delicate root tip pushes through soil particles. In the outer cells of the root cap, the golgi bodies secrete a slimy substance that lodges in the walls and eventually pass to the outside. As the cells slough off, replaced from the inside, they form a slimy lubricant that aids root tip movement through the soil. In addition, to aiding movement, the slime is a supportive medium for beneficial bacteria.

The root cap serves in an additional capacity in determining root growth direction. As the root cap has a life span of about one week, it can serve for some interesting experiments. Whether the cap sloughs off or is cut off, the root will grow in random directions, as opposed to downward, until a new root cap is formed. This lends support to the notion that the root cap functions in the perception of gravity. On the sides of the root cap amyloplasts, or plastids containing starch grains, collect facing the direction of gravitational force. In documented experiments, when the root is tipped horizontally from it's vertical growing position, the amyloplasts will reshift themselves to the "bottom" of the cells in which they are found. In a short time or 30 minutes to a few hours the root will resume growing downward. While the exact nature of this gravitational response, or gravitropism, is not fully known, there is some evidence that the calcium ions found in amyloplasts does influence the distribution of growth hormones in plant cells.

THE REGION OF CELL DIVISION

The region of cell division is the next zone in from the root cap. The root cap arises from the cells in this zone. This inverted cup-shaped region is composed of an apical meristem at it's edges. The cells divide every 12 to 36 hours at the tip of the meristem, while the ones at the base of the meristem may divide once every 200 to 500 hours. Interestingly enough, the divisions are rhythmic and peak usually twice a day around noon and midnight. In the interim the cells are not usually dividing. Most of the cells in this region are cube shaped with fairly large nuclei and few, if any, small vacuoles. As in stems as well, the apical meristem in the roots will subdivide and give rise to three meristematic areas: the protoderm, which gives rise to the epidermis; just to the inside of the protoderm, the ground meristem, which produces parenchyma cells of the cortex; and the solid looking cylinder in the center of the root, the procambium, which produces primary xylem and phloem. The central pith tissue is found in many monocots, such as grasses, but is generally not seen in mature dicot plants due to compression by the vascular cylinder.

THE REGION OF ELONGATION

This region is merged with the upper (toward the soil surface), region of the root apical meristem. It is in this region that the cells become several times their original length, and somewhat wider. The tiny vacuoles in each cell will merge and become one or two large vacuoles. In their final state, the enlarged vacuoles will account for up to 90percent or more of the cellular volume. As only the root cap and apical meristem are actually moving through the soil, no further increase in cell size occurs above the region of elongation. While the elongated portions of the root generally remain stationary for the rest of their life, if a cambium is present there may be secondary growth and an increase in root girth.

THE REGION OF MATURATION

The region of maturation is sometimes also called the region of differentiation or root-hair zone. In this region, cells mature into the various types of primary tissues. Recall that root hairs are extensions of the epidermis that serve to increase surface area and aid in absorption of water and soil nutrients. If the region of maturation is examined carefully, it would be noted that the cuticle is very thin on the root hairs and epidermal cells of roots. It is understood that any significant amount of fatty substance would interfere with the ability to absorb water, as fats are hydrophobic—or water repelling. A root in cross-section would have an epidermis, cortex, endodermis, xylem, phloem and a pericycle.

The cortex is tissue at the immediate inside of the epidermis that functions in storing food. Generally, the cortex is many cells thick and similar to the cortex of stems, with the exception of the presence of a root endodermis at the inner boundary. In stems, an endodermis is quite rare, while in roots only three species of plants are known to lack a root endodermis. The endodermis is a cylinder formed by a single layer of tightly arranged cells. The primary walls of these cells contain suberin. The waterproof suberin forms bands, called Casparian strips, around the cell walls perpendicular to the root's surface. The barrier that is formed forces all water and dissolved substances entering and leaving the central tissue core to pass through the plasma membrane or their plasmodesmata. This entire structure serves to regulate the types of minerals absorbed and transported by the root to the stems.

Next to the inside of the endodermis is a cylinder of parenchyma cells called the pericycle. The pericycle is generally one cell wide, however, it can extend for several cells depending on the plant. It is a vital tissue, as the pericycle is the point of origin for the lateral branch roots, and if it is a dicot, part of the vascular cambium. The cells in the pericycle retain their ability to divide even after they have matured. Primary xylem, which contains water conducting cells, forms at the core of the root and may or may not have observable 'branches' which extend like an 'x' to the pericycle. Primary phloem, which contains the food conducting cells, fills in the spaces between the branches of xylem. Any branch roots will usually arise in the pericycle opposite the xylem branches.

Most plants produce a fibrous root system, a taproot system, or most commonly a combination of both. However, some plants have roots with modifications that allow specific functions in addition to the absorption of water and minerals in solution.

FOOD-STORAGE ROOTS

In certain plants the roots, or part of the root system, is enlarged in order to store large quantities of starch and other carbohydrates. Sweet potatoes and yams, have extra cambial cells that develop in the xylem portion of branch roots. The cambial cells produce numerous parenchyma cells that cause the organs to swell. Starches are then stored in the swollen areas of the root. Carrots, beets and turnips have storage organs that are actually a combination of root and stem. Approximately, the top two centimeters of a carrot are actually derived from the stem.

WATER-STORAGE ROOTS

Plants that grow in particularly arid regions are known for growing structures used to retain water. Some plants in the Pumpkin Family produce huge water storing roots. The plant will then use the stored water in times or seasons of low precipitation. Some cultures will harvest the water storing root and use them for drinking water. Plants storing up to 159 pounds (72 kilograms) of water in a single major root have been found and documented.

To propagate means to produce more of oneself. Propagative root structures are a way for a plant to produce more of itself. Adventitious buds are buds that appear in unusual places. Many plants will produce these buds along the roots that grow near the surface of the ground. Suckers, or aerial stems with rootlets, will develop from these adventitious buds. The 'new' plant can be separated from the original plant and can grow independently. Some plants will produce propagative roots up to 30 feet or more away from the parent plant. This can be a nuisance for some people, while others may enjoy the propagative qualities of their cherry tree, strawberries or horseradish plants.

PNEUMATOPHORES

Pneumatophores are spongy roots that develop in most plants that grow in water. Swamps, marshes and coastal areas are good places to find plants with pneumatophores. These specialized roots account for the fact that water, even after having air bubbled through it, has less than one thirtieth of the amount of free oxygen that is found in air. Plants growing in water may require additional methods of obtaining oxygen for respiration. Pneumatophores fill that need by rising above the water surface and facilitating gas exchange.

There are many different kinds of aerial roots produced by a wide variety of plants. Orchids produce velamen roots, corn plants have prop roots, ivies have adventitious roots and vanilla orchids even have photosynthetic roots that can manufacture food. Banyan trees have aerial roots that grow down from the tree branches until they touch find the soil. In a nutshell, aerial roots are roots that are not covered by soil hence out in the air. They can facilitate climbing and various types of support as demonstrated by ivies and creeper plants.

Contractile roots are roots that pull the plant deeper into the soil. Lily bulbs are a good example, as each bulb is pulled a little further into the soil as additional contractile roots are developed each year. When a region of stable temperature is reached, the contractile roots quit pulling. Dandelions also have contractile roots, and their presence is noticeable because the lower leaves may look like they are coming right out of the ground. In reality, the roots are pulling the stem downward. The actual mechanism of contraction involves the thickening and constriction of parenchyma cells. This causes the components of xylem to spiral into a corkscrew shape. The portion of the root that contracts may lose up to two-thirds of its length within weeks.

Tropical trees may have large buttress roots at the base of the trunk. These roots add stability to the tree and give an angular look to the lower visible portion of the trunk.

Some plants, such as dodders, broomrapes and pinedrops do not have chlorophyll. They will parasitize other plants and utilize their chlorophyll and food making abilities. The parasitic mechanism involves rootlike projections called haustoria (singular haustorium). These projections develop along stems that are in contact with the host. They will penetrate the outer tissues of the host plant, and will tap into the water and food conducting tissues (xylem and phloem). Other plants with chlorphyll, such as mistletoes, will also form haustoria in order to obtain water and dissolved minerals from host plants. They are capable of producing their own food, and thus are considered to be partially parasitic.

MYCORRHIZAE

Mycorrhizae are fungal roots found in many plants. These fungal associations are important for both the plant and for the fungal and are therefore considered to be mutualistic. Essentially, the fungus will have a greater capacity for absorbing phosphorus than root hairs alone. The fungus will also grow and increase the absorption of water and other nutrients. In return, the plant provides sugars and amino acids vital to the survival of the fungus. Plants with mycorrhizae generally have less root hairs than those without. Nearly all woody trees and shrubs found in forests have fungal associations in their root systems. However, it has been demonstrated that mycorrhizae are particularly susceptible to acid rain. This may have a direct impact on forest health and maintenance.

ROOT NODULES

It is important to note that root nodules are not root knots, which are root swellings in response to worm invasions. Root nodules are beneficial bacterial colonies that are visible as small swellings in the root system. The bacteria aid the plant in fixing, or converting, atmospheric nitrogen in to a form that the plant can use. Root nodules are found extensively throughout the legume family. A nodule develops when a substance leaked into the soil by plant roots stimulates Rhizobium bacteria to produce another substance that coused root hairs to bend sharply. The bacterium may attach in the crook of the bend and then 'invade' the cell with a tubular infection thread. This thread does not penetrate the cell wall and plasma membrane. The thread, does however, grow through to the cortex which is stimulated to produce new cells that will become part of the housing for the bacterium. As the bacteria multiply and the colony grows, the nodule will swell. It is in the crook of root hairs that the nitrogen fixing takes place.

LEAVES

Leaves are highly efficient solar energy converters. They capture light energy and through the process of photosynthesis they are able to trap energy in the form of sugar molecules that are constructed from carbon dioxide and water (both found in the atmosphere). All the energy required by living organisms is ultimately dependent upon photosynthesis. Leaves are able to twist on their petioles, stalks, in order to maximize sun exposure and photosynthetic activity. Leaves are covered with a thin layer of epidermal cells which permit light to the interior of the leaf, yet protect the cells from physical damage. In addition to photosynthesis leaves are involved in other vital plant functions. Respiration is a metabolic process which produces waste products. These products are deposited outside the plant when the leaves are shed. Leaves are also important to the movement of water absorbed by the roots and transported throughout the plant. The water that reaches the leaves mostly evapourates off into the atmosphere via transpiration. Leaves are complex plant organs upon which life depends.

LEAF ARRANGEMENTS AND TYPES

There are over 275,000 different kinds of plants and most of them can be distinguished from each other by their leaves alone. As leaves originate as primordial in the buds regardless of their ultimate size and shape. When all is said and done, leaves usually consist of a stalk, the petiole and a flattened blade, the lamina, which has a network of veins also known as the vascular bundles. Some leaves have a pair of appendages called stipules at the base of their petiole. In some cases, there is no petiole or stalk, and these leaves are called sessile. Deciduous trees generally lose their leaves once a year, after the growing season. Evergreens, or conifers, usually are only functional for two to seven years.

The overall arrangement of leaves with respect to the plant stem is called phyllotaxy. Leaves may be arranged in an alternate, opposite pattern if they are attached at the same node, or a whorled pattern if three or more are attached at a node. The leaf itself may be a simple leaf, which has an undivided blade; or a compound leaf, in which the blade is divided into leaflets in various ways. A pinnately compound leaf has leaflets in pairs along a central stalk—called the rachis. A palmately compound leaf has all its leaflets attached at the same point on the end of the petiole. The leaflets of a pinnatley compound leaf are sometimes subdivided into even smaller leaflets which makes a bipinnately compound leaf. The venation, or arrangement of vascular bundles, in a leaf blade or a leaflet may be either pinnate or palmate. A pinnately veined leaf has a main vein called the midrib with secondary veins branching out from it. However, in a palmately veined leaf, several veins branch out from the base of the blade—rather than from a central midrib. Monocot plants generally have leaves with parallel venation as compared with dicots, which have branching and diverging veins. The Ginkgo tree is special in that it has no midrib or other large veins. The veins fork evenly and progressively from the base of the blade out to the opposite margin of the leaf. This arrangement is called dichotomous (branching) venation.

In cross section there are three major regions to see in the inside of a leaf: epidermis, mesophyll and veins—or vascular bundles. The epidermal layer is one cell thick and covers the entire surface of the leaf. On the lower surface of the leaf blade, the epidermis is interrupted by stomata. Which will be discussed shortly. From the top, the epidermal cells look like jigsaw puzzle pieces fit tightly together. The guard cells in the lower epidermal layer contain chloroplasts, but otherwise the epidermal cells do not have any chloroplasts and function as primary protection for the cells beneath. Most leaves have a thin covering of waxy cuticle.

STOMATA

Stomata distinguish the lower epidermis from the upper epidermis. The upper epidermis is generally uninterrupted, but the lower epidermis is perforated by numerous tiny pores called stomata. The stomata (stoma singular) are very numerous and facilitate gas exchange between the interior of the leaf and the environment. Each stoma is regulated by a pair of sausage-shaped guard cells. They, as mentioned earlier, are the only cells in the epidermis with chloroplasts for photosynthesis. The photosynthetic products in the guard cells provide the energy for the functioning of the cells. The walls of the guard cells are thickened, except for the side adjacent to the pore. The cells will expand or contract with changes in the amount of water in the cells, hence the need for energy as the water is moved into and out of the guard cells. When the guard cells are full of water the stoma pore is open and when the water is evacuated the pore is closed.

MESOPHYLL AND VEINS

The majority of photosynthesis takes place in the mesophyll between the upper and lower epidermis layers. Usually the two layers of mesophyll can be distinguished from each other. The upper region is made of cells that look like short posts in two rows. These cells are parenchyma cells and make up the palisade mesophyll tissue. It is this tissue that contains more than 80percent of the chloroplasts in the leaf.

The lower layer of mesophyll, the spongy mesophyll tissue, is composed of loosely arranged parenchyma cells with abundant air space. The lower layer also contains many chloroplasts and its loose structure allows for movement of air in from the stomata. For future reference, parenchyma tissue containing numerous chloroplasts is called chlorenchyma tissue. It is also found in the outer parts of cortex in the stems of herbaceous plants. However, in the leaf, the surfaces of the mesophyll that come into contact with the air are moist. The stomata will close if the internal moisture drops below a certain level in order to reduce drying inside the leaf.

The skeleton of a leaf are the veins, or vascular bundles. They are of various sizes and as described in the leaf arrangement section, are scattered throughout the leaf and are organized distinctly in different types of leaves. The veins are surrounded by a jacket of fibers called the bundle sheath. The sugars produced in the mesophyll are transported via the veins throughout the plant—specifically in solution in the phloem. In dicots, the veins run in all directions. In monocots, the veins are parallel and are not scattered. In addition, monocots do not have mesophyll differentiated into two layers. Instead, some will have large thin-walled buliform cells surrounding the main vein. The thin-walled cells are sensitive to water conditions and will collapse in dry conditions which causes the leaf blade to fold or roll which reduces transpiration (water loss).

Depending on the conditions where a particular plant lives, it may or may not require some specialized adaptations in order to accommodate various environmental factors: humidity, temperature, light, water, and soil conditions. We will look briefly at ten types of specialized leaves. I would suggest further research if you are interested in more detail.

1. *Shade Leaves*: In some plants, leaves with barely noticeable or unnoticeable modifications will occur right alongside those that are unmodified. Leaves in the shade tens to be thinner and have fewer hairs than those on the same tree exposed to direct light.

In addition, they are generally larger and have less defined mesophyll layers and reduced numbers of chloroplasts than their better lit counterparts.

2. *Leaves of Arid Regions*: In growing environments with extremely arid conditions, the plants will generally have thicker more leathery leaves. Their stomata are usually reduced in number and are sunken into the leaf surface in special depressions. Some may have succulent leaves or no leaves at all—where the stem takes over photosynthetic responsibilities—or they may have dense hairy coverings. In areas where the soil freezes and water resources are limited, pine trees may have modifications similar to desert plants. Including sunken stomata, thicker cuticle and a hypodermis (thick walled cells) beneath the epidermis. The compass plant is a unique example of growth set up directionally—East and West—in order to reduce moisture loss.
3. *Tendrils:* Many plants have modified leaf structures called tendrils that aid in climbing or supporting the plant's weight. Tendrils are very sensitive to contact and can be readily redirected based on touch and solid contact. Tendrils become coiled like springs and when contact with a support structure is made, the tip not only coils around it but the tip direction reverses. It needs to be noted that not all tendrils are modified leaves, tendrils of the grapevine, are modified extensions of the stem tissue.
4. *Spines, Thorns and Prickles:* Desert plants have leaves modified as spines. Water loss is correlated to surface area, so the decrease in leaf surface area consequently decreases water loss to the outside. In plants with spines, photosynthesis is generally conducted by the stem tissue. The tissue is made of sclerenchyma cells and replaces any 'normal' leaf tissues. The modifications arising in the axils of leaves are stem modifications not leaf spines, but thorns. Recall, that

the prickles of roses and raspberries are not leaves or stems, but outgrowths of the epidermal or cortex just beneath the prickle.

5. *Storage Leaves:* Succulent leaves are leaves modified to retain and store water. Water storage is permitted because of the thin-walled, non-chloroplast parenchyma cells just beneath the epidermis and to the interior of the chlorenchyma tissue. The vacuoles in the non-photosynthetic cells store the extra water resources. There are plants with succulent leaves that have a special photosynthetic process. The fleshy leaves of onions and lily bulbs store large amounts of carbohydrates which are utilized by the plant in the next growing season.
6. *Flower Pot Leaves:* The leaves of some plants, such as the Dischidia plant from tropical Australasia, develop odd pouches that become the symbiotic homes of ant colonies. The colonies carry in soil particles and add nitrogenous wastes, which the leaves collect moisture through the condensation of water vapour via the stomata. The area is a rich medium for the adventitious roots that grow down into the soil contained in the pouch—hence the flower pot function of the modified leaf.
7. *Window Leaves:* There are at least three members of the Carpetweed family in the Kalahari desert with unique adaptations to the sandy growing environment. These plants have leaves shaped like ice cream cones. The leaves are buried in the sand, leaving the transparent dime-sized tip of the leaf exposed at the surface. The transparent surface is covered with a thick epidermis and cuticle and has virtually no stomata. This arrangement allows light nearly direct access to the mesophyll with chloroplasts inside. The plant, for the most part, is buried and away from drying winds and abrasive blowing sands. There are other examples of succulent plants with window leaves.

8. *Reproductive Leaves:* Walking fern leaves produce new plants at their tips. Air plants, a succulent, have little notches along their leaf margins where new plant are produced with leaves and roots of their own. The baby plants will produce even if the parent leaf is separated from the rest of the plant.
9. *Floral Leaves (Bracts):* Bracts are found at the bases of flowers and are sometimes mistaken as petals. They compensate for small flowers or absent petals. The poinsettia 'flower' is really composed of bracts. The center cluster of tiny flowers is the main event, while the bracts do all the attracting.
10. *Insect-Trapping Leaves:* These plants are always attention grabbers and have intrigued folks for centuries. Plants that trap insects usually occur in swampy areas and bogs of tropical and temperate regions. Generally, the soil is lacking some vital ingredient for life and the plants utilize trapped insects and small organisms to fill the gap. The captured prizes are dissolved and absorbed by the plant. However, if insects are not available (i.e. a laboratory situation) the plants will develop if nutrients are given instead. The following four plants represent the four main mechanisms of capture.
    - Pitcher Plants—drowning trap
    - Sundews—sticky trap
    - Venus Flytraps—hinged trap
    - Bladderworts—underwater trapdoor trap

AUTUMN CHANGES IN LEAF COLOUR

As leaf cells break down after the growing season is over, the leaves tend to turn some shade of brown or tan due to a reaction between leaf proteins and tannins stored in the cell vacuoles. Prior to going completely tan or brown, the leaves usually demonstrate a wide variety of colours as they go through various stages of degeneration. In the chloroplasts of mature leaves are several groups of pigments such as green chlorophylls and carotenoids including yellow carotenes and pale yellox xanthophylls. These pigments play various roles in photosynthesis. The green chlorophylls are usually found in higher concentrations and during the season of active growth they are able to mask the other pigments. As the chlorophylls break down during the fall, the other colours become apparent. The breakdown of chlorophyll is not completely understood, however, it appears to be tied to the gradual reduction in day length. Anthocyanin, a common red pigment and betacyanin a second red colour may also accumulate in the cell vacuoles as fall progresses. Anthocyanins are red if the cell sap is slightly acidic and blue if the sap is more alkali (basic). Betacyanins are restricted to several plant families, including cacti and beets. While some trees demonstrate brilliant fall displays of chlorophyll breakdown, others such as birch trees have a single shade of colour in their fall leaves.

ABSCISSION

Deciduous trees and plants, the ones who lose their leaves once a year have different cycles depending on where they are at in the world. In temperate climates, the leaves generally drop in the fall in preparation for winter and new growth comes in the spring. In tropical regions, the cycle follows the cycles of wet and dry seasons. Evergreen trees do shed their leaves, however not all at once or even annually. Abscission is the process in which leaves shed; whether deciduous or evergreen.

At the base of the petiole, stalk, of each leaf there is an abscission zone. Changes that take place in this region ultimately result in the drop of leaves. Hormonal changes take place as the leaf ages and two layers of cells become differentiated. (In young leaves hormones prevent these cells from differentiating.) The cell layer closest to the stem becomes the protective layer which is usually several cells deep and suberized, or coated with a fatty suberin substance. The other layer, the separation layer, forms on the leaf side of things. The cells swell and become like jelly. The pectins in the middle lamella of the cells in the separation layer are broken down by enzymes until an external event causes the leaf to fall: this could include the force of gravity overcoming the strength of the strands of xylem holding the leaf to the petiole, thus breaking it off at the gelatinous zone, wind, rain, animals etc. The pectin breakdown begins in response to environmental conditions such as dropping temperature, lack of adequate water, decreasing day lengths, changing light intensities, or damage to the leaf.

IMPORTANCE TO HUMANS

Leaves are vital to humans. Not just for food but many medicines come from plant leaves. Tobacco products come from leaves, as do some hemp products and other textile fibers. Cocaine and aspirin are from leaves as are some insecticides. Aloe vera for the relief of burns—even x-ray burns will respond to aloe vera. Leaves are also used in floral arrangements and other products of aesthetic value. Bottom line: leaves, like stems are of great value to humans.

FLOWERS, FRUITS AND SEEDS

Flowering plants grow in a wide variety of habitats and environments. They can go from germination of a seed to a mature plant producing new seeds in as little as a month or as long as 150 years. Plants that complete their life cycle in a single season are called annuals; while biennials take two years; and perennials may take several to many years to go from germinated seed to producing new seeds. There are two major classes of flowering plants, monocots and dicots. In order to keep these two classes separate in our minds, let's take a moment and outline some of the differences between them.

DIFFERENCES BETWEEN MONOCOTS AND DICOTS

Monocots:

1. One seed leaf—cotyledon
2. Flower components in threes or multiples of three

3. Leaf veins are parallel
4. No vascular or cork cambiums
5. Vascular bundles are scattered throughout the stem
6. One aperture (thin spot) in pollen grains

Dicots:

1. Two seed leaves—cotyledon
2. Flower components in four or fives or multiples of fours or fives
3. Leaf veins are branching and networked
4. Vascular cambium present, usually cork cambium present
5. Vascular bundles are arranged in a ring in the stem
6. Three apertures in pollen grains

Dicots account for slightly under three quarters of all flowering plants. Nearly all flowering trees and shrubs are dicots as well as many annual plants. Monocots include bulb producing plants, grasses, orchids and palms. They are primarily herbaceous, meaning no secondary woody growth.

STRUCTURE OF FLOWERS

There are all sorts of flower shapes, sizes, colours and arrangements, however there are a few features that are central to all flowers regardless of their form. A flower starts as an embryonic primordium that develops into a bud and is situation as a specialized branch at the end of a stalk called the peduncle. The receptacle is a small pad-like swollen area on the very top of the peduncle. This serves as the platform for the flower parts. Whorls, which are three or more plant parts, are attached to the receptacle. The sepals are the outermost whorl and are usually green. Sometimes they are confused with leaves. They are usually three to five in number and are collectively referred to as the calyx. The second whorl of flower parts are the petals and are collectively referred to as the corolla. The corolla is usually extra-showy in order to attract pollinators. In wind-pollinated plants the corolla may be missing to maximize pollen exposure to the female flower parts. Just as the sepals in the calyx, the petals in the corolla may be fused together or separate individual units. Nestled inside the two outer whorls are the sexual organs of the flower. The stamens entail the male structures: a semi rigid filament with a sac called the anther dangling from the tip. Pollen grains develop in the anthers. Most anthers have slits or pores on the sides to accommodate pollen release. The female organs are collectively referred to as the pistil and includes: a 'landing pad' at the top called the stigma, a slender stalk like style that leads down to the swollen base called the ovary. The ovary is what will develop/ripen into a fruit.

As you might have guessed, there are names for the different ways that the flower parts are arranged with respect to the ovary. The ovary is said to be superior if the calyx and corolla are attached to the receptacle at the base of the ovary. However, if the receptacle grows up and around the ovary and the calyx and corolla are attached above it, then the ovary is said to be inferior.

Inside the ovary is an egg-shaped ovule which is held in place within the ovary by means of a short stalk. The ovule is what develops into a seed. Fruits have seeds.

Some flowers are produced all alone, while others are produced in clusters called inflorescences. An inflorescence is characterized by one peduncle with many little stalks serving individual flowers. The little stalks, in this case, are called pedicels and each stalk services one flower.

FRUITS

A fruit is a mature, or ripened, ovary that usually contains seeds. In contrast, a vegetable can consist of leaves (lettuce, cabbage), leaf petioles (celery), specialized leaves (onions), stems (white potato), stems and roots (beets), flowers and their peduncles (broccoli), flower buds (globe artichokes) and or other parts of the plant. A fruit is by definition just the ovary part of a flower, therefore all fruits come exclusively from flowering plants.

FRUIT REGIONS

A fruit, ripened ovary, has three major regions that are sometimes difficult to distinguish from each other. The outer layer, sometimes referred to as the skin, is actually called the exocarp. The mesocarp is the fleshy portion which is usually eaten when consuming fruit. The endocarp is the innermost boundary around the seed. Sometimes the endocarp is hard and stony such as a peach pit that surrounds the seed. The endocarp can also be papery as in apples, where it is barely visible in cross section. All three of these regions; the exocarp, mesocarp and endocarp, are collectively called the pericarp. The pericarp can be quite thin, as is the case with dry fruits.

Some fruits have flower parts modified or fused to the ovary at maturity. Fruits are classified according to features at maturity: fleshy, dry, split exposing seeds, non-splitting, one ovary or multiple ovaries. We will go through these various classifications and see what examples fall into the various categories.

KINDS OF FRUITS

Fleshy fruits: These fruits have a mesocarp that is at least partially fleshy at maturity.

Simple Fleshy Fruits: Fruits develop from a flower with a single pistil. The ovary may be simple, meaning derived from one modified leaf called a carpel, or compound. The ovary also may be superior or inferior and may develop into a fruit with or without other flower parts integrated.

Drupe: Drupes are simple fleshy fruits with one seed encased in a stony pit. Usually the ovary is a superior ovary with one ovule. The stone fruits—cherries, peaches, olives, apricots and almonds—are examples. Although, not readily recognized as a fleshy fruit, coconuts are drupes. The husk is the mesocarp and exocarp which is generally removed before making it to the market. The pit with the watery seed endosperm is what we see piled up at the store.

Berry: These develop from a compound ovary and usually contain multipleseeds. It is difficult to distinguish the three regions. This group is broken down further into three types of berries:

True berries are fruits with a thin skin and a pericarp. They are generally soft at maturity and usually have multiple seeds although dates and avocados are notable exceptions. Some berries have incorporated flower parts which can be seen in remnant as scars. Examples are tomatoes, grapes, peppers, blueberries, cranberries, bananas and eggplants. Note that botanically speaking raspberries, strawberries and blackberries are not berries.

Pepos are berry fruits with thick rinds. They have multiple seeds and include pumpkins, watermelons, cantaloupes and squashes.

Finally, the hesperidium is a leathery skinned berry that contains oils. Saclike outgrowths of the inner ovary wall become filled with juice as the ovary matures. All members of the Citrus Family produce hesperidium fruits.

Pome: The majority of the flesh in pomes comes from the swollen receptacle that grows up and around the ovary (inferior ovary). The seeds are encased by a leathery or papery endocarp. Apples are good examples and the apple core is the ovary with seeds, and the rest if overgrowth of receptacle. (Sometimes these fruits that are derived from more than the ovary are called accessory fruits.) Examples of pomes include: apples, pears and quinces.

Aggregate fruit: These fruits come from a single flower with multiple pistils. The individual pistils start as tiny drupes or other fruitlets, but at maturity they cluster on a single receptacle. Examples are strawberries, raspberries and blackberries.

Multiple fruit: Several of many flowers in a single inflorescence will develop into a multiple fruit. The flowers develop separately into fruitlets on their own receptacles, but at maturity they will cluster together and develop into a larger single fruit. Pineapples and figs are good examples, although the fig develops from a unique "outside in" inflorescence.

Dry fruits: Mesocarp is dry at maturity.

Dry fruits that split at maturity: distinguished by the way in which they split.

Follicle: Splits along one side, or one seam. Ex: larkspur, milkweed Legume—Splits along two sides or seams. Ex: Peas, beans, kudzu, peanuts, carob

Silique: Split along two sides or seams, the difference from legumes is that the seeds are carried on a central partition which is exposed upon splitting. Ex: Mustard Family, including broccoli, cabbage, radish, watercress

Capsule: Most common of splitting dry fruit. Composed of two carpels and split in a variety of ways: along carpel partitions, through carpel cavities, pores or via a cap that pops off to release seeds. Ex: irises, poppies, orchids, violets and snapdragons.

Dry fruits that do not split at maturity: the single seed is more or less united with the pericarp.

Achene: Seed is attached to the pericarp (husk in this case) only at the bottom, and can be separated easily. Ex: sunflower seeds (husk, plus edible seed constitutes the achene), buttercup and buckwheat.

Nut: The pericarp of nuts are generally harder than the achenes, although they are otherwise quite similar in structure. Nuts develop with a cup or cluster or bracts at their base. Ex: hazelnuts, hickory nuts and chestnuts. Note that botanically speaking most things called 'nuts' are not nuts such as peanuts (legume), coconuts, almonds, walnuts, pecans (all drupes) Brazil nuts (capsule) and pistachios (drupes). Yet another misnomer commonly accepted.

Grain: Grains are all of the Grass Family and feature a pericarp that cannot be separated from the seed. Ex. Corn, wheat, barley, rice, and oats. The grains are also called caryopses.

Samara: Pericarp extends as a wing or membrane which aids in dispersal. Usually samaras are produced in pairs, although elms and ash trees produce them singly. These are the 'helicopters' that I am certain we all have played with at one point or another. Ex: maple trees

Schizocarp: Schizocarps are the twin fruits, as at maturity the fruit dries out and breaks into two one-seeded segments. Ex: carrots, dill, parsley and anise.

FRUIT AND SEED DISPERSAL

There are a variety of methods that will get seeds from the ovary to a fertile spot to begin germinating and growing. Not all methods will work for every plant and some plants are very method specific.

DISPERSAL BY WIND

The wind can carry light seeds for miles and most seeds and fruits relying on wind dispersal have specialized adaptations. The samaras with their wings and membranes are highly ideal fruits for wind dispersal. Some fruits are too large to be carried in the air, but can be rolled along by the wind. Cottony or woolly hair type adaptations as in the Willow Family, enable better transfer of seeds via the wind. Tumbleweed plants break off and blow along in the wind, all the while dispersing seeds as it bumps along.

DISPERSAL BY ANIMALS

There are so many adaptations for the dispersal of seeds by animals that it would take a volume or two to discuss them all. Birds can carry seeds in the mud that they pick up on their feet. Seeds pass through digestive tracts and are deposited randomly by animals. Ants carry collect and carry seeds. Some seeds will not germinate unless they have passed through the acidic environment of a digestive tract. Fur and feathers can trap seeds and some seeds have burrowing type screws or hooks to ensure getting caught on something and carried along.

DISPERSAL BY WATER

Some fruits contain trapped air and are thus adapted to dispersal by water. Some pericarps are thick and spongy enough to absorb water slowly and will thus protect the tiny embryo held within. Saltwater dispersed plants generally have these type pericarps and survival requires washing up on a beach somewhere before the saltwater reaches the inside of the seed.

OTHER DISPERSAL MECHANISMS

Some fruits mechanically eject fruits, some at a violent velocity. Humans are another method of dispersal whether intentionally or not. Most countries have regulations with regards to bringing fruits and seeds into the country that may harm native species and cultivated crops.

SEEDS

We have been talking about seeds but haven't really mentioned what a seed is made of and how it becomes a mature plant.

SEED STRUCTURE

First of all dicot and monocot seeds are different. Recall that a dicot has two seed leaves in the plant embryo, while a monocot has one seed leaf. These seed leaves are called cotyledons. The cotyledons are the food storage organs and will also serve as the first leaves of the growing plant. If you look at a kidney bean—a dicot—you will notice a small white scar on the inner concave edge of the seed, which is called the hilum. The hilum is where the ovule was attached to the ovary wall—analogous to a belly button in a human. The cotyledons are attached to a tiny embryo plant contains the undeveloped leaves and meristematic tissue at one end. The embryo shoot is called the plumule and the cotyledons are attached just beneath the plumule. Above the cotyledons is the stem portion of the axis, and is called the epicotyl. The portion below the cotyledon attachment is called the hypocotyls. The plant embryo is tiny and it will be difficult to see where the stem ends and the root begins. The embryonic root is called the radicle. In some monocots, the radicle and plumule are enclosed for added protection. The tubular sheathing structures are called the coleoptile for the plumule and the coleorhiza for the radicle. At some point the embryonic shoot and root will overtake the protective structures, and the sheathing will cease growing.

GERMINATION

Germination is the start of the growing process for a plant embryo. There are a host of internal and external factors that have to be in place in order for germination to occur. Most seeds require some period of dormancy before they will germinate. This can come about by either physiological or mechanical methods or both. Some seeds can break dormancy by scarification which involves artificially cracking the seed coat. In nature, seeds may require a period of freezing and thawing in order to crack the seed coat, or passage through an acidic digestive tract. In most woody plants in temperate regions, a cold period is required before growth will commence. Some plants will absorb vast amounts of water which instigates the activity of enzymes before germination begins. When the seed is water logged oxygen may be reduced and anaerobic respiration may occur until the seed coat cracks and oxygen is admitted to the embryo. In most cases, temperature is vital to germination. Light roles in germination vary depending on the kind of plants involved.

SOILS

Where a plant grows and what resources are available to it is of vital importance to the life of a plant. The soil type and quality can be the difference between survival and termination for a plant. Soil is a very dynamic and complex portion of the earth's crust. In some places the soil is only a few centimeters thick, while in others it is hundreds of feet deep. In the grand scheme of things, soil is vital not only to us as humans, but also to the existence of nearly all living organisms. Soil contributes to the plants that grow in it, just as the growing plants give back to the soil. It is a point of dynamic exchange between the living and non-living components of the earth.

The soil as what it is today, is a result of many factors coming together: climate, parent material, local topography, vegetation, living organisms and, of course, time. All of the factors can be involved in various degrees, which is why there are many thousands of soil types.

The solid bulk of soil consists of minerals and organic matter. In between the solid particles are pore spaces, which are filled with varying amounts of air and water. The pore sizes and how they are connected within the soil bulk, determine the quality of soil aeration. Aeration refers to how water and air are held within a soil sample.

In looking closer at soils, it is important to understand that there are general regions or horizons of soil development that are usually obvious in an undisturbed area. If we found an undisturbed area, and dug down 3 to 6 feet (1 or 2 meters), we would likely find a soil profile (cross section of the horizons) of three integrated horizons.

The composition and stage of development will obviously vary widely depending on where the soil profile is taken. The top horizon is called the A horizon or topsoil. This horizon is usually 4 to 8 inches on average, again depending on where your sample is from. The A horizon is further subdivided into a darker upper portion, called the A1 horizon and a lighter lower portion A2 horizon. The A1 horizon contains the majority of the organic material out of the three integrated horizons. The next horizon is the B horizon, or the subsoil. This is usually 1 or 2 feet deep on average. The subsoil usually contains more clay, so less pore spaces, and is lighter in colour than the topsoil. The lowest horizon is called the C horizon and it could be 4 inches to 10 feet deep or it may not be present. The C horizon is called the soil parent material and it extends down to bedrock.

A1—darker upper portion of topsoil

I

I

A2—lighter lower portion of topsoil

I

I

B—subsoil

I

I

C—soil parent material

I

Bedrock

Soil Parent Material

As a quick review, soil development is first predicated on the formation of parent material. Parent material accumulates via the weathering of igneous, sedimentary and metamorphic rocks. Igneous rocks are ones from volcanic activity, sedimentary rocks are formed from glacial deposits, water or wind action, and metamorphic rocks are ones formed from the other two types after undergoing extreme pressures and heat deep within the earth.

Climate

The climate is a globally varying feature and weighs heavily upon the weathering of rocks for soil parent material. In desert or arid regions, there is little rain weathering and the soils are poorly developed. In contrast, areas of high liquid precipitation result in well developed soils. In areas where the temperature ranges widely, rocks may split and crack and cause rock breakdown. It is important to understand that the climate plays a direct role not only in the ensuring water resources to plants, but also the soil resources to plants.

Organic Composition and Living Organisms

In addition to nematode activity, the bacteria and fungi present in the soil decompose all sorts of organic material, such as leaves, dead roots, and animal carcasses. Living organisms and their organs, such as roots, produce carbon dioxide, which combines with water and forms an acid. The acidic nature results in a higher dissolving rate for the minerals present in the soil. Bacteria, fungi, nematodes, birds, ants and other burrowing insects all serve as composters for topsoil. As these organisms alter the soil through their activities, they add to it with their wastes and the decomposition of their bodies when they die.

The organic composition of soil depends on external factors, such as location and water resources. If an area is constantly wet, and oxygen is limited or lacking in the soil, the microorganism activity may be quite low and the organic content may be as high as 90percent. Thus the organic content in that soil will be quite high. On average, topsoil might be composed of 25percent air, 25percent water, 48percent minerals and 2percent organic material. Furthermore, with the exception of legumes and a few other plants, almost all of the nitrogen needed by growing plants, and most of the phosphorus and sulphur come from the decomposition of organic matter. As the decomposition progresses, the acid content increases which in turn increases the breakdown of minerals which will be carried into solution into growing plants.

Local Area Topography

Topography refers to the surface features of an area. If an area is steep soil that is weathered from parent material may quickly wash away or erode through the actions of wind, ice and water. On the other hand, if an area is flat and poorly drained, water may pool in slight depressions after it rains. This lack of drainage could result in interrupting the activities of microorganisms in the soil. Thus soil development is stopped. An ideal set of surface features would be ones that allow for soil drainage without erosion.

Mineral Composition and Soil Texture

The size of the individual soil particles is called soil texture. The three general designations are sand, silt and clay. Generally, sands and gravels are composed of many small particles bound together chemically or by a matrix of cementing material. Silt is composed of finer particles that are usually too small to see without a hand lens or microscope. Clay particles are even smaller yet. They are visible individually with a electron microscope as even a powerful light microscope will not render them visible.

Clay particles are individually called micelles. These particles are negatively charged, sheet-like and held together by chemical bonds. Because of their negative nature, they attract, trade or capture positively charged ions. The water that adheres tightly to the particle surface acts as a binding agent and a lubricant. This lends to the plastic nature of clay. Clay is also a colloid, meaning it is a suspension of particles that are larger than molecules, yet do not settle out of a fluid medium. The fluid medium is generally water. These three particles in various balances comprise the bulk of soil.

Soils are sometimes referred to as heavy and light. Light soils have a low clay content and a high sand content; while heavy soils have a high clay content and a low sand content. Clay soils have a high water content as they do not allow water to pass through between the particles, recall that water is a binding agent in clay. Sandy, or coarse-textured, soils do not retain much water. Organic matter and clay store more plant nutrients in the form of ions than sand and silt. The smaller clay and organic particles have a greater total surface area for the attachment of ions compared to an equal volume of sand and silt particles.

Soil Structure

Soil structure refers to aggregates. Aggregates are formed as a result of the arrangement and grouping of soil particles.

Sand and gravel aggregates do not demonstrate cohesion very well. In most good agricultural soils, aggregates that stick together form readily. The structure of soil, aggregates, forms when colloidal particles clump together. The clumping is usually a result of the activities of soil organisms and temperature changes resulting in freezing and thawing. Without a coating of organic matter, the individual granules will continue to clump until they are clods.

Good agricultural soils are granular with many pore spaces, occupying between 40percent and 60percent of the total soil volume. Recall how important pore spaces are to aeration and drainage. Air and water are trapped within these pores and if the pores are numerous, but small and poorly connected, as in clay soils, the restricted movement does not allow for good drainage or aeration. Furthermore, when the pores are full of water and air is not able to move in, plant roots suffer as there is not enough oxygen for root growth. Sandy soils on the other hand have their problems as well. For one, the large pores allow gravity to drain water out of the soil as soon as the pores are filled. As the water drains, the pores fill with air, which in turns speeds up nitrogen release by microorganisms. Much of the nitrogen released is lost as plants cannot utilize nitrogen that quickly.

Contrary to popular habit, watering can severely damage plants. Mainly, too much water leaches minerals and slows mineralizations processes with anaerobic conditions (due to the reduction or lack of free oxygen). Overwatering can slow the release of nitrogen, which interferes with plant growth and speed up the breakdown of nitrates.

WATER IN THE SOIL

There are three forms of water occurring in soil.

1. Hygroscopic water is unavailable to plants, because it is physically bound to soil particles.
2. Gravitational water is the water that drains from pore spaces after a rain. Plant grow can be affected if drainage is poor.

3. Capillary water is the main source of water for plant needs. This is the water held in soil pore spaces against the force of gravity. Both structure and organic material composition of the soil enable the soil to hold water in this manner. Vegetation density and type and the location of underground water tables are the determinants of how much capillary water is available to the plant.

Following rain or irrigation, gravity drains water away from the soil. The field capacity of the soil is the amount of water remaining after such a draining. This characteristic is mainly determined by the texture of the soil, however, structure and organic material content also play a role. When soil is at or near field capacity plants will readily absorb water.

In the process of soil drying, the film of water around each soil particle becomes thinner and more tightly bound to the particle. As the water film binds to the particle, the likelihood of the water entering the root is lessened. A point of no return is eventually reached, if water is not added to the plant. This point is called the permanent wilting point and refers to the soil moisture. It is at this point that the plant is unable to absorb water at a rate sufficient for its needs. The plant wilts and will die. The permanent wilting point in a clay soil is reached when the water content drops below 15percent. In sandy soils this point can be as low as 4percent. The soil water between field capacity an the permanent wilting point is called available water.

Earlier we mentioned that the breakdown of minerals lends to the acidity of the soil. It is important to understand that the pH, acidity or alkalinity, of soil affects both the soil and the plant in a variety of ways. Soil that is not balanced, that is either too acidic or too alkaline, may be toxic to roots. However, normally an unbalanced soil pH will affect nutrient availability before they will affect the plant directly. In general, when soil is acidic the nitrogen fixing abilities of plants are affected. Alkalinity will affect the availability of minerals such as copper, iron and manganese. In areas of high precipitation, soil is usually acidic because alkaline components are leached from the topsoil.

Many agricultural operations will counterbalance any soil imbalances by adding lime (compounds of calcium or magnesium) to balance acidic soil, or by adding sulphur to alkaline soils. Bacteria in the soil will convert the sulphur to sulphuric acid, which will lower the overall pH which will make the soil more acidic. Nitrogen based fertilizers may have the same effect on alkaline soils.

Chapter 5

Plant Genetic Engineering

GENETIC ENGINEERING

Genetically modified plants are created by the process of genetic engineering, which allows scientists to move genetic material between organisms with the aim of changing their characteristics. All organisms are composed of cells that contain the DNA molecule. Molecules of DNA form units of genetic information, known as genes. Each organism has a genetic blueprint made up of DNA that determines the regulatory functions of its cells and thus the characteristics that make it unique.

Prior to genetic engineering, the exchange of DNA material was possible only between individual organisms of the same species. With the advent of genetic engineering in 1972, scientists have been able to identify specific genes associated with desirable traits in one organism and transfer those genes across species boundaries into another organism. A gene from bacteria, virus, or animal may be transferred into plants to produce genetically modified plants having changed characteristics. Thus, this method allows mixing of the genetic material among species that cannot otherwise breed naturally. The success of a genetically improved plant depends on the ability to grow single modified cells into whole plants. Some plants like potato and tomato grow easily from single cell or plant tissue. Others such as corn, soy bean, and wheat are more difficult to grow.

After years of research, plant specialists have been able to apply their knowledge of genetics to improve various crops such as corn, potato, and cotton. They have to be careful to ensure that the basic characteristics of these new plants are the same as the traditional ones, except for the addition of the improved traits.

The world of biotechnology has always moved fast, and now it is moving even faster. More traits are emerging; more land than ever before is being planted with genetically modified varieties of an ever-expanding number of crops. Research efforts are being made to genetically modify most plants with a high economic value such as cereals, fruits, vegetables, and floriculture and horticulture species.

BIOLOGICAL CELL

It only takes one biological cell to create an organism. In fact, there are countless species of single celled organisms, and indeed multi-cellular organisms like ourselves.

A single cell is able to keep itself functional by owning a series of '*miniature machines*' known as *organelles*. The following list looks at some of these organelles and other characteristics typical of a fully functioning cell.

- *Mitochondrion:* An important cell organelle involved in respiration
- *Cytoplasm:* A fluid surrounding the contents of a cell and forms a vacuole
- *Golgi Apparatus:* The processing area for the creation of a glycoprotein
- *Endoplasmic Reticulum:* An important organelle heavily involved in protein synthesis.
- *Vesicles: Packages* of substances that are to be used in the cell or secreted by it.

- *Nucleus*: The "brain" of a cell containing genetic information that determines every natural process within an organism.
- *Cell Membrane:* Also known as a plasma membrane, this outer layer of a cell assists in the movement of molecules in and out the cell plays both a structural and protective role
- *Lysosomes:* Membranous sacs that contain digestive enzymes

CELL WALL

A structure that characteristically is found in plants and prokaryotes and not animals that plays a structural and protective role.

CELL SPECIALISATION

Cells can become specialised to perform a particular function within an organism, usually as part of a larger tissue consisting of many of the same cells working in tandem:

- *Nerve cells* to operate as part of the nervous system to send messages back and forth via the brain at the centre of the nerve system.
- Skin cells for waterproof protection and protection against pathogens in the open air environment.
- *Xylem* tubes to transport water around plants and to provide structural support for the plant as a whole.

Cells combine their efforts in these tissue types to perform a common cause. The task of the specialised cell will determine in what way it is going to be specialised, because different cells are suited to different purposes, as illustrated in the above list:

- Muscle cells are long and smooth in structure and their elastic nature allows these cells to perform flexible movements, just as they do in our own body's.
- Some *white blood cells* contain powerful digestive enzymes to eliminate pathogens by breaking them down to the molecular level.

- Cells at the back of the eye are sensitive to light stimuli, and thus can *interpret* differences in light intensity which can in turn be interpreted by our nervous system and brain.

Many of these cells contain organelles, though after some cells are specialised, they do not possess particular characteristics as they do not require them to be there. i.e. efficiency is the key, no resources are wasted and the resources available are put to their idyllic optimum.

THE CELL MEMBRANE

The *cell membrane*, otherwise known as the plasma membrane is a semi-permeable structure consisting mainly of *phospholipid* (fat) molecules and proteins. They are structured in a *fluid mosaic model,* where a double layer of phospholipid molecules provide a barrier accompanied by proteins.

It is present round the circumference of a cell to acts as a barrier, keeping foreign entities out the cell and its contents (like cytoplasm) firmly inside the cell.

The plasma membrane allows only selected materials to pass in and out of a cell, and is thus known as a selectively permeable membrane.

CELL TRANSPORT

There are three methods in which ions are transported through the cell membrane into the cell,

- *Active Transport*: Active transport is the transport of molecules with the active assistance of a carrier that can transport the material against a natural *concentration gradient*.
- *Passive Transport (Diffusion):* The movement of molecules from areas of high concentration (i.e. outside a cell) to areas of low concentration (i.e. within a cell) via a carrier. This process does not require energy.
- *Simple Diffusion:* The movement of molecules from areas of high concentration to areas of low

concentration in a free state. *Osmosis* of water involves this type of diffusion through a selectively permeable membrane (i.e. plasma membrane)

THE BREAKDOWN OF MATERIALS IN A CELL

In cells, sometimes it is required to breakdown more complex molecules into more simple molecules, which can then be 're-built' into what is needed by the body with these new raw materials.

'Pinocytosis' where to contents of a structure (such as bacteria) are *drank*, essentially by breaking down molecules into a drinkable form.

'Rhagocytosis' where contents are 'eaten'.

ABSORPTION AND SECRETION

Absorption is the uptake of materials from a cells' external environment. Secretion is the ejection of material.

BIOLOGICAL ENERGY - ADP AND ATP

ATP stands for Adenosine Tri-Phosphate, and is the energy used by an organism in its daily operations. It consists of an *adenosine* molecule and three inorganic *phosphates*. After a simple reaction breaking down ATP to *ADP*, the energy released from the breaking of a molecular bond is the energy we use to keep ourselves alive.

ATP TO ADP - ENERGY RELEASE

This is done by a simple process, in which one of the phosphate molecules is broken off, therefore reducing the ATP from 3 phosphates to 2, forming ADP (Adenosine Diphosphate after removing one of the phosphates {Pi}). This is commonly wrote as ADP + Pi.

When the bond connecting the phosphate is broken, *energy* is released.

While ATP is constantly being used up by the body in its biological processes, the energy supply can be bolstered by new sources of glucose being made available via eating food which is then broken down by the digestive system to smaller particles that can be utilised by the body.

On top of this, ADP is built back up into ATP so that it can be used again in its more energetic state. Although this conversion requires energy, the process produces a net gain in energy, meaning that more energy is available by re-using ADP+Pi back into ATP.

GLUCOSE AND ATP

Many ATP are needed every second by a cell, so ATP is created inside them due to the demand, and the fact that organisms like ourselves are made up of millions of cells.

Glucose, a sugar that is delivered via the bloodstream, is the product of the food you eat, and this is the molecule that is used to create ATP. Sweet foods provide a rich source of readily available glucose while other foods provide the materials needed to create glucose.

This glucose is broken down in a series of *enzyme* controlled steps that allow the release of energy to be used by the organism. This process is called respiration.

RESPIRATION AND THE CREATION OF ATP

ATP is created via respiration in both animals and plants. The difference with plants is the fact they attain their food from elsewhere.

In essence, materials are harnessed to create ATP for biological processes. The energy can be created via cell respiration. The process of respiration occurs in 3 steps (when oxygen is present):

- Glycolysis
- The Kreb's Cycle
- The Cytochrome System

CELL RESPIRATION

As mentioned in the previous page on ATP, the process of *respiration* is split into 3 distinct areas that occur at different parts of the cell. Respiration involves the *oxidation* of foodstuff (i.e. glucose) in order to create ATP.

Respiration can occur with or without oxygen, *aerobic* and *anaerobic* respiration respectively.

GLYCOLYSIS

Glycolysis occurs in the *cytoplasm* of a cell where a 6 carbon glucose molecule (the broken down food that you ate earlier) is broken down by enzymes into a 3 carbon *pyruvic acid*.

The execution of this process requires 2 ATP, and produces a net gain of 2 ATP.

The enzymes involved remove hydrogen from the glucose (oxidation) where they take these hydrogen atoms to the cytochrome system, explained soon.

In anaerobic respiration, this is where the process ends, glucose is split into 2 molecules of pyruvic acid. When oxygen is present, pyruvic is broken down into other carbon compounds in the Kreb's Cycle. When it is not present, the pyruvic acid is broken down into lactic acid (or carbon dioxide and ethanol).

THE KREB'S CYCLE

When oxygen is present, respiration can harness more ATP from a single unit of glucose. The *pyruvic acid* from the *glycolysis* stage diffuses into a cell organelle called a mitochondrion (pl. *mitochondria*). These mitochondria are sausage shaped structures that host a large surface area for the respiration to occur on.

The pyruvic acid is then subject to more enzymes which break it down into a 2 carbon compound, as seen below. The diagram illustrates the *Kreb's cycle,* consisting of three main actions

- The carbon element is in an infinite cycle where the 2 carbon compound derived from pyruvic acid binds with the 4 carbon compound that is always present in the cycle.
- CO^2 is released, where the oxygen that is present in *aerobic respiration* combines with carbon from the carbon compounds which is released as CO^2. Hence the need for animals to breath out and expel this CO^2.
- Enzymes oxidize the carbon compounds and transport the hydrogen atoms to the cytochrome system.

THE CYTOCHROME SYSTEM

The cytochrome system, also known as the hydrogen carrier system (or the *electron transport system*) are where the reduced hydrogen carriers transport hydrogen atoms from the glycolysis and Kreb's cycle stages. The cytochrome system is found in the many *cristae* of mitochondria, which are tiny stalked particles found on its outer layer.

The system contains many 'hydrogen acceptors' which hydrogen can be added to. By following the path of a hydrogen atom, we can see how the cytochrome system works:

- Some *coenzymes* from earlier stages are transferred to the next coenzymes.
- B is then oxidised, therefore the coenzyme releases the hydrogen and energy is made available.
- The released hydrogen atom binds with 2 oxygen atoms (oxygen is available in aerobic respiration) which produces water, a by-product of respiration.

The diagram illustrates this flow of hydrogen within the cytochrome system and how energy is made available by the flow of these atoms. The green circles illustrate where energy is made available via oxidation.

Overall their is a gain of 38 ATP from one molecule of glucose in aerobic respiration. The food that we eat provides glucose required in respiration. In plants, energy is also acquired via respiration, but the mechanism of delivering glucose to the respiration process is a little different.

PHOTOSYNTHESIS - PHOTOLYSIS AND CARBON FIXATION

Photosynthesis is the means that primary producers (mostly plants) can obtain energy via light energy. The energy gained FROM light can be used in various processes mentioned below for the creation of energy that the plant will need to survive and grow.

Photosynthesis is a reduction process, where hydrogen is reduced by a coenzyme. This is in contrast to respiration where glucose is oxidised.

The process is split INTO two DISTINCT areas, *photolysis* (the photochemical stage) and the *Calvin Cycle*. The diagram below gives a summary of the reaction, where light energy is used to initiate the reaction in its presence;

$$CO_2 + H_2O \rightarrow \text{glucose} + \text{oxygen}$$

PHOTOLYSIS

This part of photosynthesis occurs in the *granum* of *a chloroplast* where light is absorbed by *chlorophyll;* a type of photosynthetic pigment that converts the light to chemical energy. This reacts with water (H_2O) and splits the oxygen and hydrogen molecules apart.

From this dissection of water, the oxygen is released as a by-product while the reduced hydrogen acceptor makes its way to the second stage of photosynthesis, the Calvin cycle.

Overall, since the water is oxidised (hydrogen is removed) and energy is gained in photolysis which is required in the Calvin cycle.

THE CALVIN CYCLE

Also known as the carbon fixation stage, this part of the photosynthetic process occurs in the *stroma* of chloroplasts. The carbon made available FROM breathing in carbon dioxide enters this cycle, which is illustrated below:

Just like the *Kreb's Cycle* in respiration, a substrate is manipulated INTO various carbon compounds to produce energy. In the case of photosynthesis, the following steps occur, which create glucose for respiration FROM the carbon dioxide introduced INTO the cycle;

- Carbon FROM CO_2 enters the cycle combining with Ribulose Biphosphate (RuBP)
- A compound formed is unstable and breaks down FROM its 6 carbon nature to a 3 carbon compound called glycerate phosphate (GP)
- Energy is used to break down GP INTO triose phosphate, while a hydrogen acceptor reduces the compound therefore requiring energy
- Triose Phosphate is the end product of this, a 3 carbon compound which can double up to form glucose, which can be used in respiration.
- The cycle is completed when the leftover GP molecules are met with a carbon acceptor and then turned INTO RuBP, which is to be joined with the carbon dioxide molecules to re-begin the process.

The energy that is used up in the Calvin cycle is the energy that is made available during photolysis. The glucose that is made via GP can be used in respiration or a building block in forming *starch* and *cellulose*, materials that are commonly in demand in plants.

LIMITING FACTORS IN PHOTOSYNTHESIS

Some factors affect the rate of photosynthesis in plants, as follows

- *Temperature* plays a role in affecting the rate of photosynthesis. Enzymes involved in the photosynthetic process are directly affected by the temperature of the organism and its environment

- Light Intensity is also a *limiting factor*, if there is no sunlight, then the photolysis of water cannot occur without the light energy required.
- Carbon Dioxide concentration also plays a factor, due to the supplies of carbon dioxide required in the Calvin cycle stage.

Overall, this is how a plant produces energy which supplies a rich source of glucose for respiration and the building blocks for more complex materials. While animals get their energy FROM food, plants get their energy FROM the sun.

DNA STRUCTURE AND DNA REPLICATION

The energy required by these cells and how energy is created in order for the cell to survive.The structure, type and functions of a cell are all determined by *chromosomes* that are found in the *nucleus* of a cell. These chromosomes are composed of DNA, the acronym for deoxyribonucleic acid.

This DNA determines all the characteristics of an organism, and contains all the genetic material that makes us who we are. This information is passed on from generation to generation in a species so that the information within them can be passed on for the offspring to harness in their lifetime.

STRUCTURE OF DNA AND NUCLEOTIDES

DNA is arranged into a *double helix* structure where spirals of DNA are intertwined with one another continuously bending in on itself but never getting closer or further away.

The following diagram illustrates a *nucleotide*, the building blocks of DNA

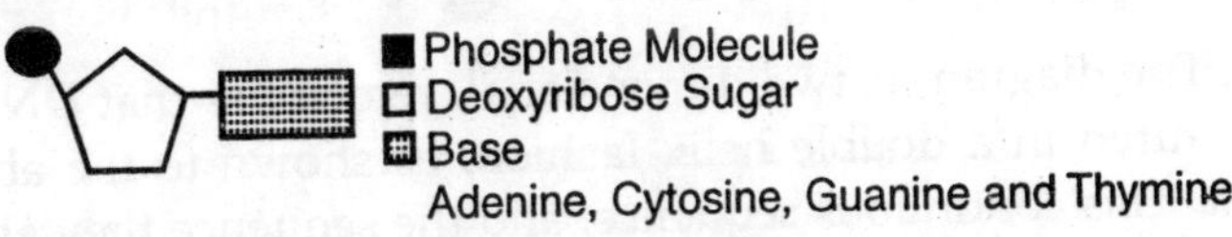

There are four different types of nucleotide possible in a DNA sequence, adenine, cytosine, guanine and thymine (can

be replaced with A, C, G and T). There are billions of these nucleotides in our genome, and with all the possible permutations; this is what makes us unique. Nucleotides are situated in adjacent pairs in the double helix nature mentioned. The following rules apply in regards to what nucleotides pair with one another.

- There are four possible types of nucleotide, adenine, cytosine, guanine and thymine.
- Thymine and adenine can only make up a base pair
- Guanine and cytosine can only make up a base pair
- Therefore, thymine and cytosine would NOT make up a base pair, as is the case with adenine and guanine.

This is illustrated in the below diagram, using correct pairings of nucleotides

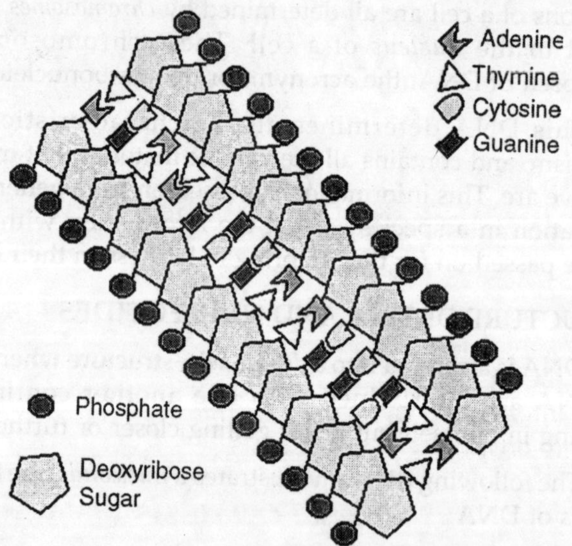

The diagram is two dimensional, remember that DNA is structured in a double helix fashion, as shown to the above right. This continuous sequence, and the sequence they are in determine an organisms' structural, physical and anatomical features.

DNA REPLICATION

Cells do not live forever, and in light of this, they must pass their genetic information on to new cells, and be able to replicate the DNA to be passed on to offspring. It is also required that fragments of DNA (genes) have to be copied to code for particular bodily function.

It is essential that the replication of it is EXACT. In order for replication to occur, the following must be available

- The actual DNA to act as an exact template
- A pool of relevant and freely available nucleotides
- A supply of the relevant enzymes to stimulate reaction
- ATP to provide energy for these reactions

When replicating, the double helix structure uncoils so that each strand of DNA can be exposed. When they uncoil, the nucleotides are exposed so that the freely available nucleotides can pair up with them.

When all nucleotides are paired up with their new partners, they re-coil into the double helix. As there are two strands of DNA involved in replication, the first double helix produces 2 copies of itself via each strand.

It is said that the replicated DNA is semi-conservative, because it possesses 50percent of the original genetic material from its parent. These 2 new copies have the exact DNA that was in the previous one. This template technique allows genetic information to be passed from cell to cell and from parents to offspring.

PROTEIN SYNTHESIS

If you have jumped straight to this page, you may wish to look at the previous page about DNA, which gives background information on protein synthesis.

As mentioned, a string of *nucleotides* represent the genetic information that makes us unique and the blueprint of who and what we are, and how we operate. Part of this genetic information is devoted to the synthesis of *proteins*, which are essential to our body and used in a variety of ways. Proteins are created from templates of information in our DNA, illustrated below:

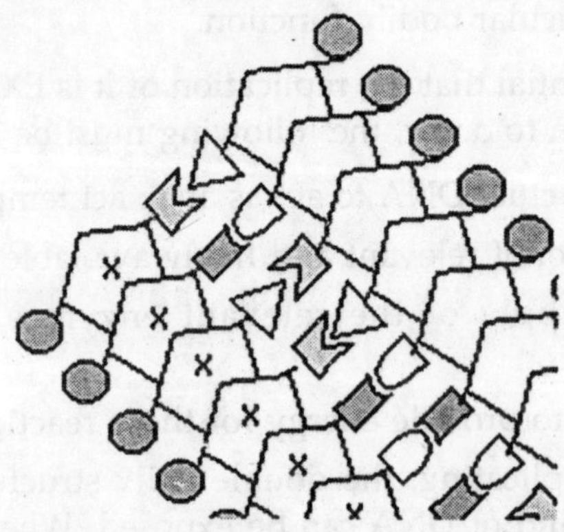

The X marked nucleotides are an example of a DNA sequence that would be used to code for a particular protein, with the sequence of these nucleotides determining which protein it is.

The sequence of these nucleotides are used to create amino acids, where chains of *amino acids* form to make a protein.

MRNA

This genetic information is found in the nucleus, though protein synthesis actually occurs in *ribosomes* found in the *cytoplasm* and on rough endoplasmic reticulum. If protein is to be synthesised, then the genetic information in the *nucleus* must be transferred to these ribosomes. This is done by *mRNA* (messenger ribonucleic acid). It is very similar to DNA, but fundamentally differs in two ways

- A base called *uracil replaces* all thymine bases in mRNA.
- The deoxyribose sugar in DNA in is *replaced* by ribose sugar in mRNA.

At the beginning of protein synthesis, just like DNA replication, the double helix structure of DNA uncoils in order for mRNA to replicate the genetic sequence responsible for the coding of a particular protein.

In the beginning, the DNA has uncoiled, allowing the mRNA to move in and transcribe (copy) the genetic information. If the code of DNA looks like this: G-G-C-A-T-T, then the mRNA would look like this C-C-G-U-A-A (remembering that uracil replaces thymine)

With the genetic information responsible for creating substances now available on the mRNA strand, the mRNA moves out of the nucleus and away from the DNA towards the ribosomes.

ROLE OF GOLGI APPARATUS AND ENDOPLASMIC RETICULUM IN PROTEIN SYNTHESIS

Continued from the previous page that introduces protein synthesis...

MRNA AND TRNA

mRNA leaves the nucleus and enters the *cytoplasm* where ribosomes can be found, the site of protein synthesis.

The mRNA strand is met in the ribosome by complimentary tRNA anticodons, which have opposing bases to that of the mRNA strand (the codons).

If the mRNA sequence is A-A-U-C-A-U, (codon) then the tRNA sequence is U-U-A-G-U-A (anticodon)

Each *tRNA* molecule consists of 3 bases, deemed an *anticodon* which compliments the opposing bases on the mRNA strand. These in turn have the *amino acid* sequence to successfully code for a particular amino acid.

Each amino acid has a certain sequence of *bases* to make it unique. Therefore, as a summary:

- The initial DNA contained a certain sequence of nucleotides
- The mRNA has a pre-determined sequence (because it is transcribed from the DNA)
- Again, as a consequence the anticodons possess a pre-determined sequence due to the mRNA
- As each amino acid corresponds to a particular anticodon, a unique amino acid sequence is created forming a protein

These amino acids (*peptides*) can combine to form a *polypeptide* chain (proteins), which are used in a variety of structures such as enzymes and hormones.

RIBOSOMES AND ROUGH ENDOPLASMIC RETICULUM (RER)

Ribosomes are the site of protein synthesis, and can occur freely in the cytoplasm though more commonly on the outer surface of rough endoplasmic reticulum. The *endoplasmic reticulum* presents a large surface area on which these ribosomes can be situated, therefore allowing protein synthesis to occur on a large scale.

Rough endoplasmic reticulum is particularly abundant in growing cells which demand a high turnover of materials in its growth. Rough ER is responsible for transporting the newly synthesised proteins to the Golgi apparatus.

THE GOLGI APPARATUS

The *Golgi apparatus* is composed of flattened fluid-filled sacs that controls the flow of molecules in a cell. This is also the case of protein. Carbohydrates are added to the protein to complete its production.

This finished product, glycoprotein, is 'pinched off' the Golgi apparatus, and is transported by a vesicle of the cell membrane. When this vesicle reaches the cell membrane, it binds to a receptor on the surface and excretes the protein, where it can then undergo its function.

PROTEIN VARIETY

As mentioned in the previous two pages investigating protein synthesis, each consists of a successive chain of amino acids. The sequence of these amino acids determine which type of protein it is. It is synthesised from a DNA strand, each DNA strand involved in protein synthesis is responsible for producing a unique protein.

TYPES OF PROTEIN

Over time and diversity of organisms, a huge amount of proteins exist and perform a unique function in the body. Primarily, their are three types of protein

- *Fibrous Proteins*: These fibre like proteins are used for structural purposes in organisms. This is because fibrous proteins are arranged in long strands and are insoluble in water. Examples of use include providing a barrier in the cell wall of plants and myosin in skeletal muscle
- *Globular Proteins*: The polypeptide chains (protein chains) in globular proteins are folded together into a knot like shape essential in the fact that are present in the following
    1. *Enzymes*: Biological catalysts, enzymes are responsible speeding up reactions in an organism
    2. *Hormones*: Hormones are chemical messengers responsible for initialising a response in organisms. Some hormones have a regulatory effect,
    3. *Antibodies*: Antibodies are used to defend the body against foreign agents e.g. bacteria, fungi and viruses. The next page investigates these.
    4. *Structural Protein*: Globular proteins form part of the cell membrane, which has a structural role as well as a role in transporting ions in and out the cell.

- *Conjugated Proteins*: Conjugated proteins are essentially globular proteins that possess non-living substances, such as the haem found in haemoglobin, which possesses iron (a non-living substance)

Therefore proteins play a vital role in many of an organisms biological processes and their organs. The following page investigates cell defence against foreign agents, where proteins are playing their role in the form of antibodies.

BIOLOGICAL VIRUSES

The prime directive of all organisms is to reproduce and survive, which is also the case for viruses, which in most cases are considered a nuisance to humans.

VIRUSES

Viruses possess both living and non-living characteristics. The unique characteristic that differentiates viruses from other organisms is the fact that they require other organisms to host themselves in order to survive, hence they are deemed *obligate parasites*.

Viruses can be spread in the following exemplar ways:

- *Airborne*: Viruses that infect their hosts from the open air
- *Blood Borne*: Transmission of the virus between organisms when infected blood enters an organisms circulatory system
- *Contamination*: Caused from the consumption of materials by organisms such as water and food which have viruses within

Therefore viruses have many means of getting transmitted from one organism to another.

CELL ASSIMILATION BY A VIRUS

Viruses are tiny micro-organisms, and due to their size and simplicity, they are unable to replicate independently. Therefore, when a virus is situated in a host, it requires the means to reproduce before it dies out without producing more viruses.

This is done by altering the genetic make up of a cell to start coding for materials required to make more viruses. By altering the cell instructions, more viruses can be produced which in turn, can affect more cells and continue their existence as a species.

The following is a step by step guide of how an example bacteriophage (a virus that infects bacteria) takes control of its host cell and reproduces itself.

- The virus approaches the bacteria and attaches itself to the cell membrane
- The tail gives the virus the means to thrust its genetic information into the bacteria
- Nucleotides from the host are 'stolen' in order for the virus to create copies of itself
- The viral DNA alters the genetic coding of the host cell to create protein coats for the newly create viral DNA strands
- The viral DNA enters its DNA coat
- The cell is swollen with many copies of the original virus and bursts, allowing the viruses to attach themselves to other nearby cells
- The process begins all over again with many more viruses attacking the hosts' cells

Without a means of defence, the host that is under attack from the virus would soon die.

BIOLOGICAL CELL DEFENCE

Organisms must find a means of defence against antigens such a viruses described on the previous page. If this was not the case, bacteria, fungi and viruses would replicate out of control inside other organisms which would most likely already be extinct.

Therefore organisms employ many types of defence to stop this happening. Means of defence can be categorised into first and second lines of defence, with the first line usually having direct contact with the external environment.

First Lines of Defence

- *Skin* is an excellent line of defence because it provides an almost impenetrable biological barrier protecting the internal environment.
- *Lysozyme* is an enzyme found in tears and saliva that has powerful digestive capabilities, and can break down foreign agents to a harmless status before they enter the body.
- The clotting of blood near open wounds prevents an open space for antigens to easily enter the organism by coagulating the blood.
- *Mucus* and *cilia* found in the nose and throat can catch foreign agents entering these open cavities then sweep them outside via coughing, sneezing and vomiting.
- The *cell wall* of plants consists of fibrous proteins which provide a barrier to potential parasites (antigens).

If these first lines of defence fail, then there are further defences found within the body to ensure that the foreign agent is eliminated.

Second Lines of Defence

Second lines of defence deal with antigens that have bypassed the first lines of defence and still remain a threat to the infected organism.

Interferons are a family of proteins that are released by a cell that is under attack by an antigen. These interferons attach themselves to *receptors* on the plasma membrane of other cells, effectively instructing it of the previous cells' situation.

This tells these neighbouring cells that an antigen is nearby and instructs them to begin coding for antiviral proteins, which upon action, defend the cell by shutting it down. In light of this, any invading antigen will not be able to replicated its DNA (or mRNA) and *protein coat* inside the cell, effectively preventing the spread of it in the organism. These antiviral proteins provide the organism with protection against a wide range of viruses.

This action brought about by interferon is a defensive measure, while *white blood cells* in the second line of defence in animals can provide a means of attacking these antigens.

One method of attacking antigens is by a method called *phagocytosis,* where the contents of the antigen are broken down by molecules called phagocytes.

These phagocytes contain digestive enzymes in their lysosomes (an organelle in phagocytes) such as lysozyme. White blood cells such as a *neutrophil* or a monocyte are capable of undergoing phagocytosis, which is illustrated below.

- The bacterium inside the cell gives out chemical messages that are picked up by the phagocyte.
- The bacteria targets the cell as a possible host and moves towards it.
- The cell is prepared for this and the bacterium becomes trapped in a vacuole that forms around it.
- The bacterium is a sitting duck that is harmless at present.
- The lysosomes detect the bacterium and the digestive enzymes inside them begin to break the bacterium down.
- The remnants of the lysosome and bacterium materials are absorbed into the cytoplasm.

The above illustrates one method of ridding an organism of an internal threat caused by an antigen. This is a non-specific response to an antigen.

PASSIVE AND ACTIVE TYPES OF IMMUNITY

The previous page investigated the role of white blood cells in phagocytosis. *White blood cells* are also responsible for *antibody* formation. Certain antibodies are synthesised in response to the presence of certain *antigens*

Specific Immune Responses

Lymphocytes are a type of white blood cell capable of producing a *specific immune response* to unique antigens. Some of these lymphocytes are capable of entrapping antigens on their surface.

When lymphocytes catch these antigens they can then begin to code for unique antibodies, structures that are capable of catching these antigens.

The lymphocytes code for a particular antibody on response to a particular antigen. The antibody that is formed will be capable of catching free antigens therefore neutralising the threat as seen below.

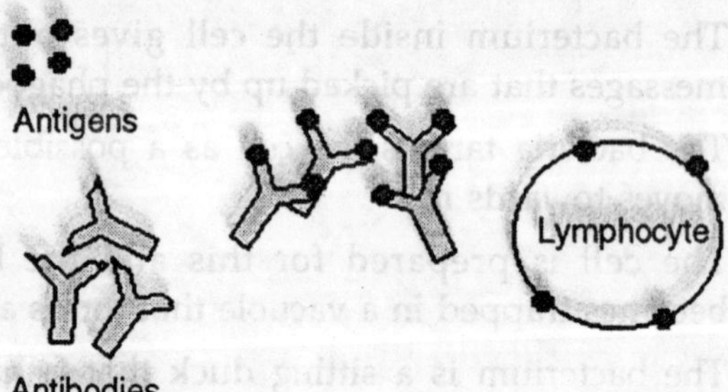

B lymphocytes (B Cells) produce free moving antibodies as above while *T lymphocytes* (T Cells) produce antibodies on their surface.

Types of Immunity

When attacked an organism has several means in which it can prepare to defend itself in event of attack.

- *Active Immunity*: *Vaccines* are used for health purposes to expose our bodies to a particular antigen. These antigens are usually killed or severely weakened to decrease their potency. After destroying these pathogens, the body stores some T cells as memory cells, due to the fact they code for a particular antigen and can be when needed. This memory in T cells can be a means of artificially acquiring immunity while a genuine attack by a pathogen is a naturally acquired type of immunity.

- *Passive Immunity*: This is where immunity to particular antigens as a result of genetic traits passed on from parents rendering the offspring immune to a particular pathogenic threat.

PLANT CELL DEFENCE

Hydrogen Peroxide

Plants release *hydrogen peroxide* in response to the presence of a fungal invasion, which attacks by piercing the cell wall of a plant and breaking it down.

This hydrogen peroxide (chemical symbol H_20_2) is a double edged sword in its defence against the antigen.

One Way: Hydrogen peroxide stops the breakdown of the cell wall

Certain pathogens will use pectinase, a digestive enzyme, to break down the cell wall barrier and invade the plant. The *pectinase* released by the fungus must be stopped. H_20_2 is involved in halting the action of this pectinase:

- The H_20_2 is created and moves to the cell wall - the site of the invasion.
- It reacts in contact with and enzyme called *peroxidase,* which promotes the breakdown of *pectinase.*
- The foreign chemical is rendered useless.
- Threat of the cell wall being compromised is removed.

Another Way: Some of the H_20_2 triggers the creation of phytoalexins

Phytoalexins are similar to the antiviral proteins previously mentioned as a secondary line of defence. Phytoalexins are a family of hormones that inhibit protein synthesis and thus "shut up shop" in the event of a pathogenic attack by halting the protein production process in our cells.

- Chemicals released by the fungus that are being used by it in its attack trigger a chemical response in the plasma membrane that makes the plant aware of the pathogens presence.
- Hydrogen peroxide from the plasma membrane triggers a chemical response to inform the nucleus of the infected cells of the current situation.
- mRNA from the nucleus is transported to ribosomes, as described in the protein synthesis section, where phytoalexins are to be produced (essentially protein synthesis coding for phytoalexins).
- The phytoalexins then take on a role similar to that of antiviral proteins, where the presence of a phytoalexin in a cell inhibits protein synthesis and therefore preventing growth of the foreign agent by removing all possible avenues of invasion for the pathogen, thus eliminating the threat.

BARRIERS USED BY PLANTS IN DEFENCE

Lignin is a strong type of molecule that provides plants with a defensive structure similar to that of fibrous proteins. It acts as a barrier and can be found in wood and is characteristically found in plants that have recently endured pathogen attack.

Callose seals off sieve plates in the plant, effectively shutting off the transport of molecules around the organism. This is done to minimise the chance of the plant transporting infectious material around its own self, and halting the movement of materials that could be used by the pathogen in aid of replicating itself.

Ethylene promotes leaf abscission, and is done to sever the plant of dead or dying plant matter. This is done to prevent the spread of infected material, therefore sacrificing infected sections of plant is more economical than taking the risk of the infection spreading.

Galls and *tannins* are created by the plant to encapsulate foreign agents found within the plant. A gall is an instance where an infected cell becomes inflamed that contains tannins. These tannins play a protective role by segregating the foreign agent and its chemicals from the rest of the plant

All of the four previous pages have illustrated means of self defence against *pathogens* (fungi, viruses and bacteria).

PLANT TRANSFORMATION

Genetic engineering of plants is much easier than that of animals. There are several reasons for this:

1. There is a natural transformation system for plants (the bacterium *Agrobacterium tumefaciens*),
2. Plant tissue can redifferentiate (a transformed piece of leaf may be regenerated to a whole plant), and
3. Plant transformation and regeneration are relatively easy for a variety of plants.

The soil bacterium *Agrobacterium tumefaciens* ("tumefaciens" meaning tumor-making) can infect wounded plant tissue, transferring a large plasmid, the Ti plasmid, to the plant cell. Part of the Ti (tumor-inducing) plasmid apparently randomly integrates into the chromosome of the plant.

The integrated part of the plasmid contains genes for the synthesis of

1. Food for the bacterium, and
2. *Plant hormones*: Genes from the Ti plasmid that are integrated in the plant chromosome are expressed at high levels in the plant.

Overproduction of the plant hormones leads to continuous growth of the transformed cells, causing plant tumors. Rapid, cancerous growth of the transformed plant tissue obviously is advantageous to the bacterium: more food gets produced.

The Ti plasmid has been genetically modified ("disarmed") by deleting the genes involved in the production of bacterial food and of plant hormones, and inserting a gene that can be used as a selectable marker. Selectable marker genes generally are coding for proteins involved in breakdown of antibiotics, such as kanamycin. Any gene of interest can be inserted into the Ti plasmid as well. In principle, one can thus transform any plant tissue, and select transformants by screening for antibiotic resistance.

However, unfortunately, there are some complications:

1. It has proven difficult to transform some monocots (grasses, etc.) by *Agrobacterium*, and
2. Regeneration of plants from tissue culture or leaf discs is not always possible.

A number of genetically engineered plant varieties have been developed. Traits that have been introduced by transformation include herbicide resistance, increased virus tolerance, or decreased sensitivity to insect or pathogen attack. Traditionally, most of such genetically engineered plants were tobacco, petunia, or similar species with a relatively limited agricultural application. However, during the past decade it now has become possible to transform major staples such as corn and rice and to regenerate them to a fertile plant. Increasingly, the transformation procedures used do not depend on *Agrobacterium tumefaciens*. Instead, DNA can be delivered into the cells by small, μm-sized tungsten or gold bullets coated with the DNA. The bullets are fired from a device that works similar to a shotgun. The modernized device uses a sudden change in pressure of He gas to propel the particles, but the principle of "shooting" the DNA into the cell remains the same. This DNA-delivery device is nicknamed "gene gun", and has been shown to work for DNA delivery into chloroplasts as well. Over the last several years, use of the "gene gun" has become a very common method to transform plants, and has been shown to be applicable to virtually all species investigated. Transformation of rice by this method is now routine. This is a very important development as rice is the most important crop in the world in terms of the number of people critically dependent on it for a major part of their diet.

Another method to get foreign genes into cereals is by electroporation: a jolt of electricity is used to puncture self-repairing holes in protoplasts (i.e., the cell without the cell wall), and DNA can get in through these holes. However, it is often very difficult to regenerate fertile plants from protoplasts of cereals. Nonetheless, significant advances in overcoming these practical difficulties have been made over the years. Now even transgenic trees have been created: the gene for a coat protein of the plum pox virus has been introduced into apricot. The plum pox virus leads to the feared Sharka disease, for which there is no cure. The resulting transgenic tree shows a markedly decreased sensitivity to this virus. The reason why continuous exposure of the tree to the viral coat protein leads to tolerance against viral infection is not yet understood, however.

Thus, now there are a number of different techniques to introduce foreign genes into plants. Essentially all major crop plants can be (and have been or are being) genetically engineered, the procedures are now routine and the frequency of success is very high. Even though genetically engineered crops are more costly than the usual ones, they have been rather readily accepted by US farmers provided that tangible benefits can be demonstrated. However, it is questionable whether the farmer in poorer countries can come up with the funds to "try out" and use the new crops. Another issue in this respect is how genetically engineered crops are perceived by the consumer. Even though in the US there is little resistance to such crops as long as the products can be shown to be safe and advantageous, in other countries genetically modified foods are received poorly by the consumer. It is unlikely that there is a rationally sound basis for this rather hostile reaction of the consumer, as most of the crops are the result of human manipulation and may have been treated with harmful herbicides and pesticides. Time and education will need to be invested to provide consumers and consumer advocates with a balanced opinion on the acceptability of the origin of their foods. One area of particular concern for some people is the lack of labeling of genetically engineered foods, and legislation may be introduced to address this issue. On the other hand, as so many plants (soybean, corn, etc.) are genetically modified and the nature of the genetic modification is not necessarily easy to explain, it may be simpler to label those foods that are guaranteed free of "genetically modified organisms" or their products. However, keep in mind that essentially all agricultural products have been genetically modified by traditional breeding, so it may be difficult to define what is actually free of genetically modified organisms.

APPLICATIONS OF PLANT GENETIC ENGINEERING

"PHARMING" AND "PLANTIBODIES"

An increasingly viable option is the production of highly valuable enzymes by plants and animals. In addition to production of human proteins in these organisms, other valuable proteins that are currently produced by microorganisms could very well be produced by higher organisms instead. Animal and plant "bioreactors" in some respects may be superior to recombinant bacterial systems, because eukaryotes glycosylate proteins. Whereas the glycosylation pattern may be species-specific, appropriate glycosylation is often required for protein function. Production through these organismal systems may also be cheaper than cell fermentation techniques.

Two examples of production of human proteins in plants include the production of human serum albumin in transgenic tobacco and potato, and production of human insulin by tobacco. In both cases, the produced protein appears to be fully effective in humans. Unfortunately, however, one cannot raise his/her insulin level by eating transgenic tobacco leaves, as the protein in most cases will be broken down to amino acids before it reaches the blood stream. Therefore, in these cases one cannot escape the practice of protein isolation and purification before transgenic leaves are converted into drugs.

Also antibodies are being produced in plants. Initially, the antibody's light and heavy chains were produced in different plants. But a subsequent cross of these two varieties resulted in progeny carrying assembled and functional antibodies.

REVERSIBLE MALE STERILITY IN PLANTS

As has been indicated earlier, heterozygous individuals often are healthier and stronger than homozygous ones. The only way to guarantee heterozygoiscity in plants is to make sure self-pollination cannot occur. For most crop plants it was very tedious or practically impossible to exclude selfing. To exclude self-pollination, it would be good to introduce male sterility in plants: progeny from such plants are then expected to be 100percent heterozygous (assuming they were pollinated with pollen from an unrelated variety). To introduce male sterility, a promoter was identified that was turned on exclusively in tapetum cells (a tissue around the pollen sac that is essential for pollen production). This promoter then was linked up to a gene coding for a bacterial ribonuclease (named barnase). This ribonuclease selectively chops up ribonucleic acids. The promoter/ribonuclease construct was then introduced into plants (canola, tobacco, you name it). Because the promoter allows expression only in tapetum cells, the gene construct disrupts only development of the tapetal tissue and its end product, pollen. Plants transformed with this construct were male-sterile but otherwise normal.

Although male-sterile plants are valuable for hybrid seed production, they have limited value when it comes to crop production. Fertility must be restored to crops such as wheat, rice, and tomato, in which the seed or fruit is the harvested product. Fortunately, the ribonuclease is inhibited very much by a simple protein, named barstar. One can thus cross the male-sterile plant with a male-fertile variety in which the gene for barstar has been introduced, and the result is progeny with viable pollen and restored fertility.

A closely related approach has been criticized as "terminator technology" as it is seen by its critics as a way for companies to protect and enforce their patents.

ANTISENSE RNA

Antisense RNA refers to nucleotide strands that are produced in a cell and that are complementary to a particular mRNA. Antisense RNA can be produced, by inverting the coding region of a gene with respect to its promoter. The antisense RNA can hybridize with its corresponding mRNA, making it double-stranded. The double-stranded mRNA no longer can be recognized by the protein-synthesizing machinery (the ribosomes), and thus expression of this mRNA is suppressed. Also, in many systems double-stranded mRNA is very unstable and is broken down quickly. Thus, one can inactivate specific genes while not interfering with others.

Antisense approaches already are used to protect plants from damage by plant viruses. Reversal of a gene from bean yellow mosaic virus (BYMV), and putting it into tobacco under a reasonably strong promoter, has led to a tobacco variety that is quite resistant to BYMV. A similar approach is used to transfer viral resistance to other plants. This finding is of significance, in that currently no effective, environmentally friendly methods exist to control many plant viruses.

Very related to this approach is the RNAi (RNA interference) approach. This is very useful for both agriculture and medicine, and the first examples of practical applications of this RNAi technology are appearing. As with any new technology, the initial pilot projects are sort of pedantic. However, more exciting application possibilities abound. Obviously, RNAi technology provides an excellent approach for reverse genetics in eukaryotes. With this method one can turn off genes.

AGRICULTURAL APPLICATIONS IN DEVELOPING COUNTRIES

Perhaps indicative of the large potential and relative ease of genetic engineering, developing countries (particularly China) are progressing rapidly in development and application of genetically engineered crops. Some have gone into commercial production well ahead of similar crops in the US. In China, tomatoes that have been engineered for improved virus resistance have been on the market since late 1992. There are two main reasons for the more rapid commercialization of bioengineered crops in the developing world: (1) less tight governmental approval mechanisms, and (2) hungrier populations. While in developed countries the main value of biotechnological applications may be to reduce production costs, in the developing world a main factor is the production of more food. Indeed, genetic engineering applications seem to be pretty successful to cut down on pathogen-induced losses. Genetic modification of papaya plants (expression of the ringspot virus coat protein in the plant) protects very well against the very destructive ringspot virus.

Chapter 6

DNA, RNA and Protein Synthesis

STRUCTURE AND FUNCTION OF DNA

DNA molecules are incredibly long, but also very thin. One DNA molecule from the chromosome of a mammal may be about 1 m long when unraveled. However, it has to fit in a nucleus of some 5-6 orders of magnitude smaller and is folded up in chromosomes in a highly organized manner. DNA is a linear polymer that is composed of four different building blocks, the nucleotides. It is in the sequence of the nucleotides in the polymers where the genetic information carried by chromosomes is located. Each nucleotide is composed of three parts:

1. A nitrogenous base known as purine (adenine (A) and guanine (G)) or pyrimidine (cytosine (C) and thymine (T));
2. A sugar, deoxyribose; and
3. A phosphate group.

The nitrogenous base determines the identity of the nucleotide, and individual nucleotides are often referred to by their base (A, C, G, or T). One DNA strand can be up to several hundred million nucleotides in length. T can form a hydrogen bond with A, and C with G; two DNA strands wind together in an antiparallel fashion in a double-helix.

Inside the cell, the DNA acts like an "instruction manual": in its sequence, it provides all the information needed to function, but the actual work of translating the information into a medium that can be used directly by the cell is done by RNA, ribonucleic acid. The structural difference with DNA is that RNA contains a -OH group both at the 2' and 3' position of the ribose ring, whereas DNA (which stands, in fact, for deoxy-RNA) lacks such a hydroxy group at the 2' position of the ribose. The same bases can be attached to the ribose group in RNA as occur in DNA, with the exception that in RNA thymine does not occur, and is replaced by uracil, which has an H-group instead of a methyl group at the C-5 position of the pyrimidine. The RNA has three functions:

1. It serves as the messenger that tells the cell (the ribosomes) what protein to make (messenger RNA; mRNA);
2. It serves as part of the structure of the ribosome, the protein/RNA complex that synthesizes proteins according to the information presented by the mRNA (ribosomal RNA; rRNA); and
3. It functions to bring amino acids (the constituents of the proteins) to the ribosome when a specific amino acid "is called for" by the information on the mRNA to be put in into the protein that is being synthesized; this RNA is called transfer RNA (tRNA).

An important point of emphasis should be that all vegetative cells of one organism contain the same genetic information. Upon division, each daughter cell obtains an "exact" copy of the DNA of the parent. However, the specific genes that are expressed at specific times may be very different between different tissues. These differences in gene expression allow for the regulation of development of the organism, and for the development of different tissues. For the most part, DNA-binding proteins (encoded by the DNA) play an important role in the regulation of expression of genes encoded on the DNA. A very important "chicken-and-egg" problem.

RNAS

The messenger RNA (mRNA) serves as an intermediate between DNA and protein. Parts of the DNA are "transcribed" into transcripts (single-stranded RNA molecules) that are processed to mRNA. In prokaryotes the transcript generally does not need to be processed, and can serve as mRNA right away. Transcription starts at a specific site on the DNA called a promoter. Each gene or operon has its own promoter(s). Transcription ends at a terminator sequence on the DNA. The transcripts usually are 300-50,000 nucleotides long, and contain the information to make protein. In eukaryotes (organisms with cells containing a nucleus; in fact, any higher organism) generally the transcripts needs to be processed before they can serve as a blueprint for a protein. The processing involves the removal of intervening sequences (introns) in the gene. The introns may be anywhere between 50 and 10,000 nucleotides in length. The coding regions of the mRNA are called exons. There may be up to 100 introns in a single gene. The introns are spliced out by small ribonucleoprotein particles (consisting of RNA and protein), which appear to pull the two ends of the intron together. However, there are also introns that splice out without the need of a protein: the RNA sequence itself appears to contain sufficient information to know where to splice out the intron. In addition to the removal of introns, a poly-A sequence is added to the 3′ end of the transcript. The processed transcript is the mRNA, and the information in the mRNA can be used to be "translated" into a protein of specific sequence. However, in prokaryotes introns are rare and mRNA generally does not get processed before translation.

The intron splicing process provides an opportunity to increase the amount of usable genetic information without increasing the genome size of the organism: Alternative splicing of a particular transcript can occur. Alternative splicing means that introns may be recognized in different ways in different molecules of the same primary transcript, and the result is that one gene can give rise to different mRNAs and thereby to different proteins. Note that this process is largely limited to eukaryotes as introns in prokaryotes are rare.

Ribosomal RNAs (rRNAs) are essential components of an important part of the protein synthesis machinery: the ribosomes. In addition to rRNA, there are some 70 different proteins in a ribosome. There are hundreds of copies of rRNA genes per genome, thus making the production of lots of rRNA possible. There are four different rRNAs, each with a different size. Each ribosome contains one molecule of each of the four rRNA types. In prokaryotes, ribosomes bind to the mRNA close to the translation start site. This ribosome binding site is referred to as the Shine-Dalgarno sequence or as the ribosome recognition element. In eukaryotes, ribosomes bind at the 5' end of the mRNA and scan down the mRNA until they encounter a suitable start codon.

Transfer RNA (tRNA) carries amino acids to the ribosomes, to enable the ribosomes to put this amino acid on the protein that is being synthesized as an elongating chain of amino acid residues, using the information on the mRNA to "know" which amino acid should be put on next. For each kind of amino acid, there is a specific tRNA that will recognize the amino acid and transport it to the protein that is being synthesized, and tag it on to the protein once the information on the mRNA calls for it.

All tRNAs have the same general shape, sort of resembling a clover leaf. Parts of the molecule fold back in characteristic loops, which are held in shape by nucleotide-pairing between different areas of the molecule. There are two parts of the tRNA that are of particular importance: the aminoacyl attachment site and the anticodon. The aminoacyl attachment site is the site at which the amino acid is attached to the tRNA molecule. Each type of tRNA specifically binds only one type of amino acid. The anticodon (three bases) of the tRNA base-pairs with the appropriate mRNA codon at the mRNA-ribosome complex. This temporarily binds the tRNA to the mRNA, allowing the amino acid carried by the tRNA to be incorporated into the polypeptide in its proper place. Thus, the sequence of the codon (three bases) in the mRNA dictates the amino acid to be put in in the protein at a specific site. The "dictionary" of codons coding for amino acids is called the genetic code. The three codons for which there is no matching tRNA (UAA, UGA, and UAG) serve as "stop-translation" signals at which the ribosome falls off.

PROTEIN SYNTHESIS

After having discussed DNA and the various RNAs, the stage has been set for protein synthesis. The basic reaction of protein synthesis is the controlled formation of a peptide bond between two amino acids. This reaction is repeated many times, as each amino acid in turn is added to the growing polypeptide. Protein synthesis starts when the mRNA binds to a small ribosomal subunit near a AUG sequence in the mRNA. The AUG codon is called start codon, since it codes for the first amino acid (a methionine) to be made of the protein. The AUG codon base-pairs with the anticodon of tRNA carrying methionine. A large ribosomal subunit binds to the complex, and the reactions of protein synthesis itself can begin. The aminoacyl-tRNA to be called for next is determined by the next codon (the next three bases) on the mRNA. Each amino acid is coded for by one or more (up to six) codons. Of course, it would be more straightforward to have each amino acid coded for by only one codon, but nature appears to have chosen a more complex route. The reason for this in part is that there are 20 different amino acids, and 4x4x4=64 different combinations possible in a codon. When the ribosome reaches one of the three codons for which there is no matching tRNA, the ribosome falls off and the synthesized protein is released.

RNA EDITING

Over the last several years, it has become obvious that the sequence present in DNA does not always dictate literally the sequence of the protein. In a number of instances "RNA editing" has been observed (particularly in the small genomes present in mitochondria and chloroplasts), in which transcripts are chemically modified by enzymes before translation takes place. Thus, the DNA sequence in such cases does not precisely correlate with the sequence of the gene product (the protein). One thus needs to compare sequences from DNA and protein (or from DNA and processed RNA) if one suspects that RNA editing can occur. The function of RNA editing has not been elucidated yet.

AMPLIFICATION OF DNA

DNA SEQUENCING

With the development of automated DNA sequencers any sequence can be obtained readily and rapidly. However, it should be kept in mind that cloning and DNA isolation (required for DNA sequencing by any means) is also labour-intensive. Automation at that level is available as well, but remains expensive. Larger sequencing laboratories employ automation at the DNA sequencing, sample preparation, and interpretational levels, and the amount of new sequence information that has become available over the past few years has grown almost exponentially. The science of interpreting DNA sequence information and accessing it in ways that are as efficient as possible is named bioinformatics.

Genomic sequencing currently is carried out by breaking the genome into small pieces (1,000-3,000 nucleotides), cloning them, and sequencing each piece individually. The entire sequence is then put together by overlapping the sequences of all pieces. Particularly for larger genomes the pasting together is challenging in view of the occurrence of repeats in the sequence, but with sufficient controls a reliable result can be obtained. Currently, a large laboratory specializing in sequencing can obtain a complete genomic sequence of a bacterium in just a day or so. However, functional assignment of genes etc. takes much longer as this requires brain power and sometimes detailed experimentation for verification.

POLYMERASE CHAIN REACTION (PCR)

The practical applications of PCR are quite amazing. With this method sufficient DNA can be prepared from single hairs or from blood stains to do Southern blots or DNA sequencing. This of course is of importance for forensic applications. For paleontologists and archeologists, the application is a little different: Sufficient DNA can be isolated from a mammoth that froze to death in Siberia many millennia ago, from a Bronze-Age hunter found recently in the ice of a glacier in the Alps, from mummies in Egypt, from 8,000-year-old human brain tissue, from magnolia leaves encapsulated in Miocene shale, or from 11,000-year-old bison bones, to amplify to quantities sufficient for sequencing. However, the amount of DNA needed for sequencing is becoming smaller: In 2005, rather degenerated DNA isolated from a long extinct cave bear was sequenced without amplification.

The record of DNA recovery from ancient organisms currently is 40 million years: reasonably intact DNA has been recovered from a termite that was embedded in amber some two dozen million years after the dinosaurs became extinct. From sequencing of specific genes, evolutionary relationships between the old (sub)species and current representatives can be deduced. Even more reminescent of "Jurassic Park", ancient bacterial spores from the intestines of a 25-40 million year old bee that was preserved in amber reportedly have been revived and cultured. The resulting bacterial culture is most closely related to *Bacillus sphaericus*.

An interesting PCR application involves the identification of the remains of the Romanovs, the Russian czar family killed in 1918 by soldiers of the Bolshevik army as ordered by Stalin. The killing of czar Nicholas II and his family is one of the haunting tales of the Russian Revolution. According to "legend", the bodies of the czar family and that of servants and the family doctor were buried in a shallow road-side grave near Jekatarinenburg (formerly Sverdlovsk) after the truck on which their corpses were carried broke down; according to the same "legend", concentrated sulfuric acid was poured over the corpses to make positive identification impossible (or so they thought), and bones were broken and scattered through the grave. After the end of the Soviet era, the location of the putative grave was disclosed, and indeed some badly decomposed and partially broken human bones were found. But how can one positively identify this grave as belonging to the czar and his household? Best start with bone (that's all they had to go by anyway). One gram of sulfuric acid-treated bone may yield some 10-50 pg (1 pg = 10^{-12} g) of DNA, enough to do PCR. Both tandem-repeat chromosomal and hypervariable mitochondrial sequences were amplified. Note that mitochondrial DNA is inherited maternally only. Sex determination of bones was done by amplification of the gene for amelogenin that is common to X and Y chromosomes, and that yields different PCR sizes depending on whether it is on an X or a Y chromosome.

A key player in the positive identification of the bones as belonging to the czar family was Prince Philip, husband of the British Queen. He is the great-grandson, by maternal lineage, of the mother of the czar's spouse. The mitochondrial DNA sequence of Prince Philip matched up with DNA from four reconstructed skeletons, one of an adult and three of children, but not with any of the other skeletons. This implies the adult skeleton is that of the czar's spouse, and the other three belong to three of her five children. Comparing PCR sequences of nuclear DNA from the children's bones with that of the remaining skeletons in the grave, the czar's skeleton could be identified. Also, four of the nine skeletons are not related to each other or to one of the other five. These four skeletons presumably are of three servants and of the family doctor. The skeletons of two of the children, presumably those of Alexei (the tsarevitz) and Anastasia, one of the daughters, were not in the grave. They either survived or were burned and buried separately. With the current knowledge on the nuclear and mitochondrial genome of the czar's family, it is easy to develop a DNA fingerprint of the missing children. So far, none of the claims have proven to be true; however, any proven heirs may await what is left of the family fortune.

GENETICS AND EVOLUTION

MEIOSIS - THE GENETICS OF REPRODUCTION

The genetic information found in *DNA* is essential in creating all the characteristics of an organism. This remains the case when passing genetic information to offspring, that can occur via a process called *meiosis* where four *haploid* cells are created from their *diploid* parent cell.

For a species to survive, and genetic information to be preserved and passed on, reproduction must occur. This can be done by passing on the information found in the chromosomes via the gametes that are created in meiosis.

CHROMOSOME COMPLEMENT

Humans are diploid creatures, meaning that each of the *chromosomes* in our body are paired up with another.

Haploid cells possess only one set of a chromosome. A diploid human cell possesses 46 chromosomes and a gamete created by a human is haploid possesses 23 chromosomes.

Tetraploid organisms possess more than 3 sets of a particular chromosome.

REPRODUCTION

Reproduction occurs in humans with the fusion of two haploid cells (*gametes*) that create a *zygote*. The nuclei of both these cells fuse, bringing together half the genetic information from the parents into one new cell, that is now genetically different from both its parents.

This increases *genetic diversity*, as half of the genetic content from each of the parents brings about unique offspring, which possesses a unique *genome* presenting unique characteristics. Meiosis as a process can increase genetic variation in many ways, explained soon.

THE PROCESS OF MEIOSIS

The process of meiosis essentially involves two cycles of division, involving a gamete mother cell (diploid cell) dividing and then dividing again to form 4 haploid cells. These can be subdivided into four distinct phases which are a continuous process

1st Division

- Prophase - *Homologous chromosomes* in the nucleus begin to pair up with one another and then split into chromatids (one half of a chromosome) where crossing over can occur. Crossing offer can increase genetic variation.
- Metaphase - Chromosomes line up at the equator of the cell, where the sequence of the chromosomes lined up is at random, through *chance,* increasing genetic variation via independent assortment.
- Anaphase - The homologous chromosomes move to opposing poles from the equator
- Telophase - A new nuclei forms near each pole alongside its new chromosome compliment.

At this stage two haploid cells have been created from the original diploid cell of the parent.

2nd Division

- Prophase II - The nuclear membrane disappears and the second meiotic division is initiated.
- Metaphase II - Pairs of chromatids line up at the equator
- Anaphase II - Each of these chromatid pairs move away from the equator to the poles via *spindle fibres*
- Telophase II - Four new haploid gametes are created that will fuse with the gametes of the opposite sex to create a zygote.

Overall, this process of meiosis creates gametes to pass genetic information from parents to offspring, continuing the family tree and the species as a whole.

INDEPENDENT ASSORTMENT AND CROSSING OVER

The previous page investigates the process of meiosis, where 4 *haploid* gametes are created from the parent cell. Half the genetic information from a parent is present in these haploids, which fuse with gametes of the opposite sex to create a *zygote,* with a complete chromosome compliment that will create offspring after prolonged growth.

The process of *meiosis* increases *genetic diversity* in a species. The sex organs which produce the haploid gametes are the site of many occurrences where genetic information is exchanged or manipulated.

INDEPENDENT ASSORTMENT OF CHROMOSOMES

Alleles for a particular *phenotype* determine what characteristic an organism will express, as with the following example where

- Chromosome 1 contains an allele for blonde hair
- Chromosome 2 contains an allele for brown hair
- Chromosome 3 contains an allele for blue eyes
- Chromosome 4 contains an allele for brown eyes

The top assortment to the left produces 2 blonde hair/blue eyes gametes while the below produces 2 brown hair/brown eyes gametes

The top assortment on the right produces 2 blonde hair/ brown eyes gametes while the below produces 2 brown hair/ blue eyes gametes

The above indicates that even though the two homologous chromosomes contain the same genetic information, the assortment of the chromosomes (the order they lie in) can determine what genetic information is present in each of the 4 gametes produced. With 23 chromosomes in a human gamete, their are 2^{23} combinations (8388608 combinations)

CROSSING OVER

During meiosis, when *homologous chromosomes* are paired together, there are points along the chromosomes that make contact with the other pair. This point of contact is deemed the *chiasmata,* and can allow the exchange of genetic information between chromosomes. This further increases *genetic variation.*

CROSSING OVER AND GENETIC DIVERSITY

Gregor Mendel, an Austrian monk, is most famous in this field for his study of the *phenotype* of pea plants, including the shape of the peas on the pea plants.

GREGOR MENDEL'S WORK

Mendel's goal was to have a firm scientific basis on the relationship of genetic information passed on from parents to offspring. In light of this he focused on how plant offspring acquired the phenotype of their seeds.

The plants that were used in the experiment had to be *true breeding,* i.e. those plants with round seeds must have had parents with round seeds, who in turn had parents producing round seeds etc. This is done to increase the accuracy of results.

After successfully producing two generations from these true breeding plants, the following was evident

- The first generation of plants produced all had a round seed phenotype.
- When these first generation plants were crossed, a ratio of 3 round seeds averaged every 1 wrinkled seed.
- The ratio of 3:1 was not exact, though this is because of the randomness of the processes that are executed to produce these plants. Independent assortment is completely random, as are mutations, therefore variable results occur producing a sampling error.
- Due to the scale of the experiment done by Gregor Mendel, the sampling error was smaller than that of a smaller scale experiment.

Mendel successfully hypothesised that the reason for this trend in phenotypes from generation to generation was down to the fact that genetic information was being passed on from their parents.

The fact that round seeds appeared more frequently than wrinkled seeds is due to round seeds being the dominant phenotype, which when present effectively 'masks' the phenotype of the recessive (wrinkled seed) gene.

DOMINANT AND RECESSIVE ALLELES

- All plant seeds produced in the first generation were round
- 3 out of every 4 plant seeds produced in the second generation were round

  The parents, one possessing wrinkled seeds the other possessing round were crossed together, for some reason in the first and second generation the presence of the round seed gene in offspring superceded the presence of the wrinkled seed. This is called dominance.

MENDEL'S LAW & MENDELIAN GENETICS

Mendel's First Law

"The alleles of a gene exist in pairs but when gametes are formed, the members of each pair pass into different gametes. Thus each gamete contains only one allele of each gene."

INCOMPLETE DOMINANCE

When a particular gene possesses both *dominant* and recessive alleles, it is possible for *incomplete dominance* to occur, where the organism at hand expresses a phenotype morphed by the expression of both the dominant and recessive alleles.

In essence, *heterozygous* (possessing opposing alleles Rr) organisms derived from *homozygous* (possessing the same alleles RR or rr) are created, they possess a phenotype different to that of both their parents.

MULTIPLE ALLELES

Diploid organisms naturally have a maximum of 2 alleles for each gene expressing a particular characteristic, one deriving from each parent. In some cases, however, more than two types of allele can code for a particular characteristic, as is the case of genetic coding for blood type in humans. Their are up to 6 possible genotypes that code for the four blood groups, A, B, AB and O.

EXAMPLE OF A CROSS

The following dihybrid cross involves two true breeding pea plants, where two factors are looked at, the shape of the seed and the colour of the seed.

MENDELIAN GENETICS

The past few pages have elaborated on the work of Gregor Mendel and how his work has paved the way to predicting the characteristics of offspring. However, a degree of randomness is involved, when involving factors such as independent assortment during meiosis and the possibility of genetic mutations.

CHROMOSOME MUTATIONS

It is natures intention that the exact genetic information from both parents will be seen in the offspring's *DNA* in the the critical stages of fertilisation. However, it is possible for this genetic information to mutate, which in most cases, can result in fatal or negative consequencies in the outcome of the new ogranism.

Non-Disjunction and Down's Syndrome

One well known example of mutation is *non-disjunction*. Non-disjunction is when the spindle fibres fail to seperate during meiosis, resulting in gametes with one extra chromosome and other gametes lacking a chromosome.

If this non-disjunction occurs in chromosome 21 of a human egg cell, a condition called *Down's syndrome* occurs.

This is because their cells possess 47 chromosomes as opposed to the normal chromosome compliment in humans of 46.

The fundamental structure of a chromosome is subject to mutation, which will most likely occur during crossing over at meiosis. There are a number of ways in which the chromosome structure can change, as indicated below, which will detrimentally change the genotype and phenotype of the organism. However, if the chromosome mutation effects an essential part of DNA, it is possible that the mutation will abort the offspring before it has the chance of being born.

The following indicates types of chromosome mutation where whole genes are moved:

Deletion of a Gene

As the name implies, genes of a chromosome are permanently lost as they become unattached to the centromere and are lost forever

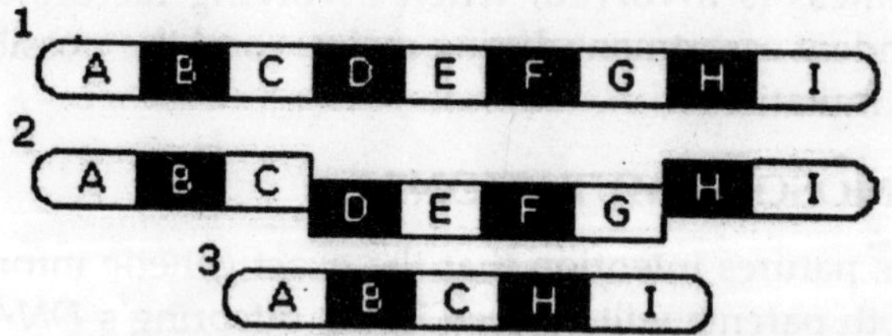

- Normal chromosome before mutation.
- Genes not attached to centromere become loose and lost forever.
- New chromosome lacks certain genes which may prove fatal depending.

Duplication of Genes

In this mutation, the mutants genes are displayed twice on the same chromosome due to duplication of these genes. This can prove to be an advantageous mutation as no genetic information is lost or altered and new genes are gained

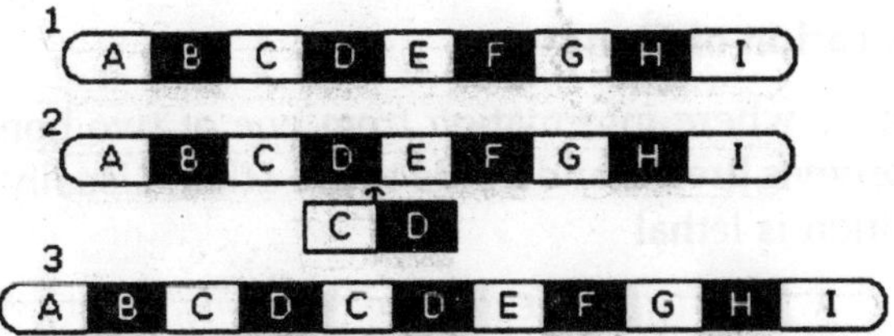

- Normal chromosome before mutation
- Genes from the homologous chromosome are copied and inserted into the genetic sequence
- New chromosome possesses all its initial genes plus a duplicated one, which is usually harmless

GENETIC MUTATIONS

This page continues from the previous page investigating genetic mutations.

Inversion of Genes

This is where the order of a particular order of genes are reversed as seen below

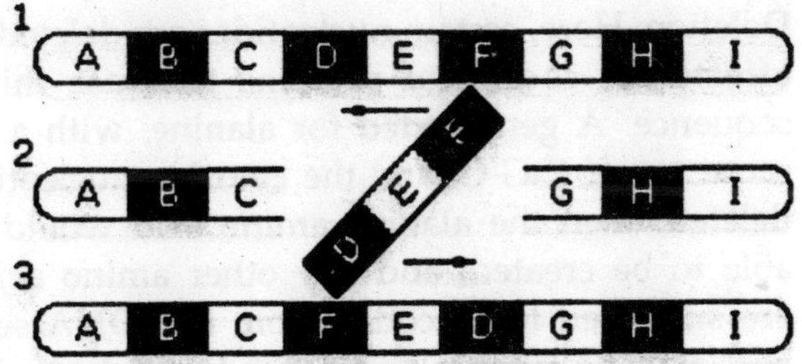

- Normal chromosome un-altered
- The connection between genes break and the sequence of these genes are reversed
- The new sequence may not be viable to produce an organism, depending on which genes are reversed. Advantageous characteristics from this mutation are also possible

Translocation of Genes

This is where information from one of two homologous chromosomes breaks and binds to the other. Usually this sort of mutation is lethal

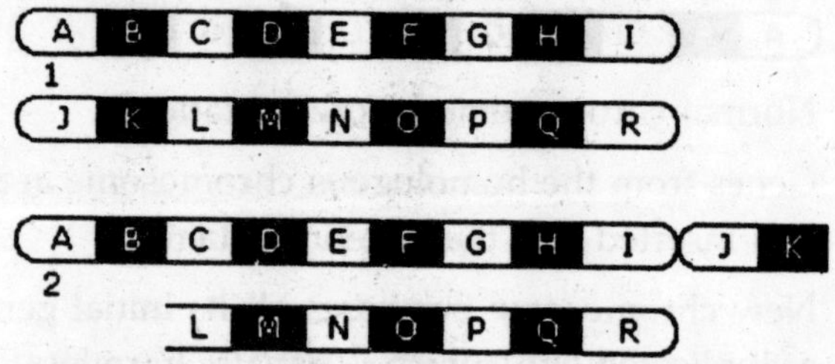

- An un-altered pair of homologous chromosomes
- Translocation of genes has resulted in some genes from one of the chromosomes attaching to the opposing chromosome

Alteration of a DNA Sequence

Mutation have investigated changes at the chromosome level. The sequence of nucleotides on a DNA sequence are also susceptible to mutation.

- Deletion: Here, certain nucleotides are deleted, which affects the coding of proteins that use this DNA sequence. A gene coded for alanine, with a genetic sequence of C-G-G, and the cytosine nucleotide was deleted, then the alanine amino acid would not be able to be created, and any other amino acids that are supposed to be coded from this DNA sequence will also be unable to be produced because each successive nucleotide after the deleted nucleotide will be out of place.
- Insertion: Similar to the effects of deletion, where a nucleotide is inserted into a genetic sequence and therefore alters the chain thereafter. This alteration of a nucleotide sequence is known as frameshift
- Inversion: Where a particular nucleotide sequence is reversed, and is not as serious as the above

mutations. This is because the nucleotides that have been reversed in order only affect a small portion of the sequence at large

- Substitution: A certain nucleotide is replaced with another, which will affect any amino acid to be synthesised from this sequence due to this change. If the gene is essential, i.e. for the coding of haemoglobin then the effects are serious, and organisms in this instance suffer from a condition called sickle cell anaemia.

All of the genetic mutations looked at through the last 2 pages more or less have a negative impact and are undesired, however, in some cases they can prove advantageous.

Genetic mutations increase genetic diversity and therefore have an important part to play. They are also the reason many people inherit diseases.

MUTATION FREQUENCY AND POLYPLOIDY

The previous two pages have investigated mutations, and this page continues with more information related to genetic mutations.

Polyploidy

Humans are *diploid* creatures, meaning for every chromosome in our body, there is another one to match it. Read the following

- Haploid creatures have one of each chromosome
- Diploid creatures have two of each chromosome
- Triploid creatures have three of each chromosome
- Polyploid creatures have three or more of each chromosome
- They can be represented by n where n equals haploid, 2n equals diploid and so on.

It is possible for a species, particularly plant species, to produce offspring that contains more chromosomes than its parent. This can be a result of non-disjunction, where normally a diploid parent would produce diploid offspring, but in the case of non-disjunction in one of the parents, produces a polyploid.

In the case of *triploids,* although the creation of particular triploids in species is possible, they cannot reproduce themselves because of the inability to pair homologous chromosomes at meiosis, therefore preventing the formation of gametes.

Polyploidy is responsible for the creation of thousands of species in today's planet, and will continue to do so. It is also responsible for increasing genetic diversity and producing species showing an increase in size, vigour and an increased resistance to disease.

Mutation Frequency

This page and the previous two have investigated the different ways that mutations arise, and the following elaborates on the ways in which mutations are instigated.

Barring all external *factors,* mutations occur very rarely, and are rarely expressed because many forms of mutation are expressed by a recessive allele.

However there are many mutagenic agents that artificially increase the rate of mutations in an organism. The following are some factors that increase genetic mutations in organisms

- Members of species in a particular geographic area or ethnic origin are more susceptible to mutations
- High dosages of X-Rays or ultraviolet light can increase the likeliness of a mutation
- Radioactive substances increase the rate of mutations exponentially

As mentioned previously, genetic mutations are a source of new variation in a species because it physically alters the sequence of nucleotides in a given sequence, therefore altering the genome in a unique way.

THEORY OF NATURAL SELECTION

In the 19th century, a man called *Charles Darwin*, a biologist from England, set off on the ship HMS Beagle to investigate species of the island.

After spending time on the islands, he soon developed a theory that would contradict the creation of man and imply that all species derived from common ancestors through a process called natural selection. Natural selection is considered to be the biggest factor resulting in the diversity of species and their genomes. The principles of Darwin's work and his theory are stated below.

- One of the prime motives for all species is to reproduce and survive, passing on the genetic information of the species from *generation* to generation. When species do this they tend to produce more offspring than the *environment* can support.
- The lack of resources to nourish these individuals places pressure on the size of the species *population*, and the lack of *resources* means increased competition and as a consequence, some organisms will not survive.
- The organisms who die as a consequence of this competition were not totally random, Darwin found that those organisms more suited to their environment were more likely to survive.
- This resulted in the well known phrase *survival of the fittest*, where the organisms most suited to their environment had more chance of survival if the species falls upon hard times. (This phrase if often associated with Darwin, though on closer inspection Herbert Spencer puts the phrase in a more accurate historical context.)
- Those organisms who are better suited to their environment exhibit desirable characteristics, which

is a consequence of their genome being more suitable to begin with.

This 'weeding out' of less suited organisms and the reward of survival to those better suited led Darwin to deduce that organisms had evolved over time, where the most desirable characteristics of a species are favoured and those organisms who exhibit them survive to pass their *genes* on.

As a consequence of this, a changing environment would mean different characteristics would be favourable in a changing environment. Darwin believed that organisms had 'evolved' to suit their environments, and occupy an *ecological niche* where they would be best suited to their environment and therefore have the best chance of survival.

As the above indicates, those alleles of a species that are favoured in the environment will become more frequent in the genomes of the species, due to the organisms higher likeliness of surviving as part of the species at large

DARWIN'S FINCHES & NATURAL SELECTION

Darwin's finches are an excellent example of the way in which species' gene pools have adapted in order for long term survival via their offspring. The Darwin's Finches diagram below illustrates the way the finch has adapted to take advantage of feeding in different ecological niche's.

Their beaks have evolved over time to be best suited to their function. The finches who eat grubs have a thin extended beak to poke into holes in the ground and extract the grubs. Finches who eat buds and fruit would be less successful at doing this, while their claw like beaks can grind down their food and thus give them a selective advantage in circumstances where buds are the only real food source for finches.

Industrial Melanism

Polymorphism pertains to the existence of two distinctly different groups of a species that still belong to the same species. Alleles for these organisms over time are governed by the theory of natural selection, and over this time the genetic differences between groups in different environments soon become apparent, as in the case of industrial melanism.

Industrial melanism occurs in a species called the peppered moth, where the occurrence has become of more frequent occurrence since the beginning of the industrial age. The following argument elaborates the basis of principles involved in natural selection as far as industrial melanism is concerned.

- *Pollution,* which is more common in today's world since the industrial age causes a change in environment, particularly in the 1800's when soot would collect on the sides of buildings from chimneys and industries and make them a darker colour.
- The resultant effect was that the peppered moth, which had a light appearance was more visible against the darker backgrounds of sooty buildings.
- This meant that *predators* of the peppered moth could find them more easily as they are more visible against a dark background.
- Due to mutations, a new strain of peppered moth came to existence, where their phenotype was darker than that of the white peppered moth.
- This meant that these new, darker peppered moths were once again harder to track down by their prey in environments where industry has taken its toll.
- In this instance, natural selection would favour the darker moths in polluted environments and the whiter moths in the lesser polluted environments due to their ability to merge in with their environmental colours and lessen the chances of them being prone to a predator.

Sickle Cell Trait

Consider this argument of natural selection in the case of sickle cell trait, a genetic defect common in Africa.

- Sickle cell trait is a situation that occurs in the presence of a recessive *allele* coding for *haemoglobin,* a substance in the blood responsible for the transport of gases like oxygen. The presence of the allele is either partially expressed recessively (sickle cell), or fully expressed by a complete recessive expression which results in full blown *anaemia*. If this particular allele is dominant, no sickle cell trait is expressed in the phenotype.
- The above occurrences in the case of a recessive allele result in structural defects of red blood cells, severely reducing the organisms capacity to uptake oxygen.
- It was pointed out that in Africa, there is a high frequency of this mutation, where cases of *malaria* were high.
- A substantiated link was made noting those who suffer sickle cell trait or anaemia were immune to the effects of malaria.
- This is yet again natural selection at work. Although sickle cell trait or anaemia are not advantageous characteristics on their own, they prove to be advantageous in areas where malaria proves to be a greater threat to preserving the genome (i.e. surviving).
- The *incomplete dominance* of this genetic expression proves favourable either way.

This is how science has understood natural selection since the first studies involving Darwin. In the 21st century, humans selectively breed species to create hybrid species possessing the best genes of both parents via a process known as selective breeding.

SELECTIVE BREEDING

As mentioned in previous pages, scientists from the past harnessing the knowledge of genetics has resulted in many scientific breakthroughs and uses of this knowledge. Most notably, Gregor Mendel's studies into Monohybrid and Dihybrid crossing and Charles Darwin's study of evolution and natural selection has meant that humans have learnt to actively manipulate the *phenotype* of offspring by selective breeding in animals and plants.

Breeders of animals and plants in today's world are looking to produce organisms that will possess desirable characteristics, such as high crop yields, resistance to disease, high growth rate and many other phenotypical characteristics that will benefit the organism and species in the long term.

This is usually done by crossing two members of the same species which possess *dominant* alleles for particular genes, such as long life and quick *metabolism* in one organism crossed with another organism possessing genes for fast growth and high yield. Since both these organisms have dominant genes for these desirable characteristics, when they are crossed they will produce at least some offspring that will show ALL of these desirable characteristics. When such a cross occurs, the offspring is termed a *hybrid,* produced from two genetically dissimilar parents which usually produces offspring with more desirable qualities. Breeders continuously track which characteristics are possessed by each organism so when the breeding season comes once again, they can selectively breed the organisms to produce more favourable qualities in the offspring.

The offspring will become *heterozygous,* meaning the allele for each characteristic will possess one dominant and one recessive gene. Most professional breeders have a true breeding cross (ie AAbb with AAbb) so that they will produce a *gene bank* of these qualities that can be crossed with aaBB to produce heterozygous offspring. This way the dominant features are retained in the first breeding group and can be passed on to offspring in the second instance.

This process of selecting parents is called artificial selection or selective breeding, and poses no threat to nature from man manipulating the the course of nature. It has allowed our species to increase the efficiency of the animals and plants we breed, such as increasing milk yield from cows by continuously breeding selected cows with one another to produce a hybrid.

Inbreeding Depression

However, while it is an advantage both to the species and to humans to produce these desirable qualities that may benefit the organisms in question, continuous in-breeding and selective breeding of particular genes runs the risk of losing some of the other genes from the gene pool altogether, which is irreversible. This is called in-breeding depression, where the exclusivity of the advantageous genes mean that some other less desirable genes are phased out. In the long term, it is more advantageous for organisms to remain heterozygous;

· *Genetic diversity* means the *gene pool* of a species is prepared for a wide range of scenarios such as food shortage or an epidemic of disease. Some genes in some organisms may provide the organism with *immunity* against the disease or an ability to go long periods of time without food. If continuous in-breeding has occurred in a species, some of these genes may have been phased out due to the breeder wanting other more desirable genes to be present in their crop....

Genetic diversity in the long term is reduced, because many organisms end up with similar *genomes* due breeding with each other constantly. In normal circumstances, this process would be random, and would produce more variable offspring

With the above facts in hand, breeders need to produce more heterozygous offspring to ensure the long term welfare of the species they are breeding and their livelihood. The most important thing here is to preserve the genetic diversity of a species, and preferably keep the gene pool of a species as diverse as possible.

Humans have realised the above dangers, and instead of harnessing and exhausting natures reserves, we have learned to preserve their genetic information for their long term survival and our own well-being. One species becoming extinct can knock the balance of an ecosystem and have a detrimental knock on effect.

With this in mind, humans have gene banks to preserve the genetic information in the case of extinction, and nurture species that are at dangerously low population levels.

Ironically, the human interference that has disrupted so many species can now provide a means of placing genes into organisms, therefore preparing them for the above hypothetical scenarios such as an epidemic of disease. Genetic engineering would provide the means of allowing organisms to suit their environment without the trial and error over time that comes from natural selection.

GENETIC ENGINEERING ADVANTAGES & DISADVANTAGES

During the latter stage stages of the 20th century, man harnessed the power of the atom, and not long after, soon realised the power of genes. Genetic engineering is going to become a very mainstream part of our lives sooner or later, because there are so many possibilities advantages (and disadvantages) involved. Here are just some of the advantages:

- Disease could be prevented by detecting people/plants/animals that are genetically prone to certain *hereditary* diseases, and preparing for the inevitable. Also, infectious diseases can be treated by implanting genes that code for antiviral proteins specific to each antigen.
- Animals and plants can be 'tailor made' to show desirable characteristics. Genes could also be manipulated in trees to absorb more CO^2 and reduce the threat of global warming.
- Genetic Engineering could increase genetic diversity, and produce more variant alleles which could also be crossed over and implanted into other species. It

is possible to alter the genetics of wheat plants to grow insulin.

Of course there are two sides to the coin, here are some possible eventualities and disadvantages:

- Nature is an extremely complex inter-related chain consisting of many species linked in the food chain. Some scientists believe that introducing genetically modified genes may have an irreversible effect with consequences yet unknown.
- Genetic engineering borderlines on many moral issues, particularly involving religion, which questions whether man has the right to manipulate the laws and course of nature.

Genetic engineering may be one of the greatest breakthroughs in recent history alongside the discovery of the atom and space flight, however, with the above eventualities and facts above in hand, governments have produced legislation to control what sort of experiments are done involving genetic engineering. In the UK there are strict laws prohibiting any experiments involving the cloning of humans. However, over the years here are some of the experimental 'breakthroughs' made possible by genetic engineering.

- At the Roslin Institute in Scotland, scientists successfully cloned an exact copy of a sheep, named 'Dolly'. This was the first successful cloning of an animal, and most likely the first occurrence of two organisms being genetically identical. Note: Recently the sheep's health has deteriorated detrimentally
- Scientists successfully manipulated the genetic sequence of a rat to grow a human ear on its back. (Unusual, but for the purpose of reproducing human organs for medical purposes)
- Most controversially, and maybe due to more liberal laws, an American scientist is currently conducting tests to clone himself.

Genetic engineering has been impossible until recent times due to the complex and microscopic nature of DNA and its component nucleotides. Through progressive studies, more and more in this area is being made possible.

For us to understand chromosomes and DNA more clearly, they can be mapped for future reference. More simplistic organisms such as fruit fly (Drosophila) have been chromosome mapped due to their simplistic nature meaning they will require less genes to operate. At present, a task named the Human Genome Project is mapping the human genome, and should be completed in the next ten years.

The process of genetic engineering involves splicing an area of a chromosome, a gene, that controls a certain characteristic of the body. The enzyme endonuclease is used to split a DNA sequence and split the gene from the rest of the chromosome. This gene may be programmed to produce an antiviral protein. This gene is removed and can be placed into another organism. It can be placed into a bacteria, where it is sealed into the DNA chain using ligase. When the chromosome is once again sealed, the bacteria is now effectively re-programmed to replicate this new antiviral protein. The bacteria can continue to live a healthy life, though genetic engineering and human intervention has actively manipulated what the bacteria actually is. No doubt there are advantages and disadvantages, and this whole subject area will become more prominent over time.

THE GENE POOL AND SPECIATION

We have already discussed some of the reasons why the genetics of a species can change over a long period of time. Charles Darwin's The Origin of Species went into great depth in this matter, and providing substance into the theory of evolution. The key thing to remember about evolution is that it favours more preferable genes in the gene pool, and over time, these preferable characteristics become more exclusive in the gene pool.

This next section rounds up all the factors that can alter the make up of a gene pool

Natural Selection

Natural selection will favour genes that are more suited to their environment and become more exclusive in the gene pool over time in such an environment. Different genes will become more exclusive when the environment changes, or the species migrate.

Mutation

Mutations are random occurrences which change the genome of an organism. They greatly increase genetic diversity, where advantageous mutations are favoured by natural selection and disadvantageous ones are phased out.

Gene Migration

Occurs after *genetic drift*, where two groups of a species become separated and therefore cannot reproduce. The gene pool of these groups differ over time. If these two groups can once again meet up and reproduce, their genetic differences can be merged within the single group and increase *genetic diversity*.

Non-Random Mating

Can cause in-breeding depression by continuous inbreeding, non-random mating is also known as selective breeding, where the breakthroughs of Mendelian genetics have allowed us to predetermine what genes are present in offspring. As advantageous genes are desired by the breeder, some of the less 'popular' genes are lost due to this random mating, therefore decreasing genetic diversity.

It is important for a species to have a large gene pool, because in the event of danger, some alleles will allow the species to survive and reproduce to produce a larger and more variant gene pool. An extremely contagious disease may threaten 99percent of a species, though the remaining 1percent may possess an allele that provides them with resistance to the disease. If this allele was not present in the population, then chances are the entire population would be wiped out

Genetic Drift

Sometimes, species can be split into groups, usually as a result of a geographical factor preventing the two groups from contacting one another. This means that the two groups are unable to reproduce with one another.

As the two groups now live in different environments, natural selection will favour slightly different genes in each of the groups that will favour them in their particular environment. Over time, the difference in gene pool between the two groups can be quite dramatic.

If these two groups once again become re-united, gene migration occurs. If they remain seperated, their genetic differences become greater which can result in the formation of a new species.

Speciation

When the genetic differences become so great, it can come to the stage where the two groups can no longer reproduce with one another. This results in the formation of a new species (because organisms who are capable of reproducing but cant reproduce with a member of the same species are deemed another species).

Long ago, the land of Earth was all on one continent. Over time these continents separated, with members of the same species drifting away on each continent. Over time, speciation has occurred. This is evident by looking at the marsupials of Australia, who have been isolated from other mammals in their ancestral line, and therefore have many differences to that of mammals on other continents.

Adaptive Radiation

Adaptive radiation is the slow change of *genotype* and *phenotype* of a species from its common ancestor, meaning that species with a common ancestor become more diversified over time. The next page investigates adaptive radiation.

ADAPTIVE RADIATION

When *Charles Darwin* was in the Galapagos islands, one of the first things he noticed is the variety of finches that existed on each of the islands. All in all, there were many different species of finch which differed in beak shape and overall size. This is adaptive radiation and natural selection at work.

The Marsupials

The marsupial mammals occupy Australia, and dither from placental mammals because they bear their young inside a pouch. Long ago, the land mass of Earth consisted of one single continent, Pangaea, where all animals existed. When this continent fragmented into smaller continents, the geographical barrier meant the mammals of the time could no longer reproduce with one another. From here on in, the gene pool of these groups of mammals would become increasingly different until they could no longer reproduce with one another. This occurrence applies to the marsupial family of mammals that occupy Australia.

All the marsupials in present day Australia would have evolved from one common ancestor. However, over time and via natural selection, the many marsupial species (i.e. kangaroo and koala) have occupied their own ecological niche and adapted accordingly. Kangaroo's have long powerful legs to cover the wide area of land that they occupy while the koala's smaller structure and more centralised centre of gravity allow them to climb trees and obtain the eucalyptus that they feed on.

Humans, the most evolved species on the planet, have also underwent many changes over time from our ancestors.

- Humans are bipedal, meaning we walk on two of our limbs. The other limbs, our arms and subsequent hands, have adapted to do precision tasks such as typing or tying shoe laces, common tasks that we perform in our own environment.

- The amount of melanin in our skin is representative of the environment we live in, i.e. dark skinned people occupy hotter climates.
- The structure of our body gives an indication of the temperature in our climate also. Humans who have produced offspring that successfully live in a cold environment tend to be broader and smaller in stature while hotter environments are occupied by thinner taller humans.

It is worth noting that if two groups of humans were totally isolated over time, it is entirely possible that a different species could spawn from own, because the two groups would be adapting to their own environment over time. This would mean changes in the gene pool in each of the groups, to the point where the two groups could not reproduce with another and produce fertile offspring due to these genetic differences. At this point, with any species, a new one has begun.

MEIOSIS AND ALTERNATION OF GENERATIONS

MITOSIS

The cell cycle contains the process in which cells are either dividing or in between divisions. Cells that are not actively dividing are said to be in interphase, which has three distinct periods of intense activity that precedes the division of the nucleus, or mitosis. The division of the rest of the cell occurs as an end result of mitosis and this process occurs in regions of active cell division, called meristems. Meristems.

Mitosis is a process within the cell cycle that is divided into four phases which we will sum up here:

1. Prophase: The chromosomes and their usual two-stranded nature becomes apparent, the nuclear envelope breaks down.

2. Metaphase: The chromosomes become aligned at the equator of the cell. A spindle composed of spindle fibers is developed and some attach to the chromosomes at their centromere.
3. Anaphase: The sister chromatids of each chromosome, that is now called the daughter chromosomes, separate lengthwise and each group of daughter chromosomes migrates to the opposite ends of the cell.
4. Telophase: The groups of daughter chromosomes are grouped within a developing nuclear envelope which makes them separate nuclei. A wall forms between the two sets of daughter chromosomes thus creating two daughter cells.

In plants, as the cell wall is developing, droplets or vesicles of pectin merge forming a cell plate that eventually will become the middle lamella of the new cell wall.

The key feature of mitosis is that the daughter cells have the same chromosome number and are otherwise identical to the parent cell.

MEIOSIS INTRODUCTION

Mitosis, as just reviewed above, is a process by which a cell can reproduce itself and the number of chromosomes and the nature of the DNA will be identical to the original parent cell. Very few species will grow or live indefinitely, so there must be some way to ensure the continuity of the species. Reproduction is the only way a species can be perpetuated, without perpetuation the species will become extinct. Reproduction can occur in several ways as vegetative propagation, such as in the development of runners in strawberry plants, or by special cells called vegetative spores which are products of mitosis. In these processes, the 'offspring' have identical cells and identical chromosomes to the parent cells and thus the processes are called asexual reproduction—a means without, so without sex reproduction. Most plants, however, will undergo sexual reproduction which involves the production and recombining of sex cells called gametes. In flowering and cone-bearing plants this involves the production of seeds. The gametes produced are male and female, and are called sperm cells and egg cells, correspondingly. When the gametes combine together, the cells fuse and form a single cell called a zygote. It is the zygote that will go on to become the plant embryo and eventually a mature, adult plant.

However, in thinking about this process, what would happen if both gametes had the same number of chromosomes as the rest of the cells in the organism? When they fused to become a zygote, they would have two times the number of chromosomes as the rest of the cells in the organism. The number of chromosomes would increase exponentially through the generations if this occurred. This is where meiosis comes in to play. Meiosis is the process by which gametes, sex cells, are formed. It is unique because gametes have exactly half of the total number of chromosomes as the rest of the cells in the parent organism. When two gametes, each with half the number of chromosomes, get together they are able to restore the chromosome number to the same as the rest of the cells in the parent organism. When the zygote develops into a plant embryo and eventually a mature plant, it will have the exact number of chromosome specific to the species. Note that the processes and steps in meiosis are very similar to mitosis, so make certain you have a good understanding of mitosis so that you will be able to compare the two processes.

Before we get into the nitty-gritty of meiosis, keep in mind that all living cells have two sets of chromosomes—one from a male and one set from a female parent. The genes in the chromosomes may control the same characteristics but in contrasting ways: genes for plant height, genes for plant colour, genes for fruit colour, etc—the female gamete might code for short plants, while the male gamete might code for tall plants. That is more of a genetics topic though. But you should know that the chromosomes that code for the same characteristics are called homologous chromosomes.

PHASES OF MEIOSIS:

The end result of one round of meiosis will be four cells with half the number of chromosomes as the parent cell. The daughter cells are rarely, if ever, identical to each other or the parent cell depending on the organism involved. There are two successive divisions in meiosis, which in plants occur without a pause. Mitosis takes roughly 24 hours, while meiosis takes up to two weeks. In some organisms, meiosis takes weeks or years depending on the organism.

Division I –Reduction division–

The chromosome number is reduced to half the parent cell chromosome number. End result of division one is two cells.

Prophase I

1. Chromosomes coil, becoming shorter and thicker, the two-stranded nature becomes apparent, two strands are called a chromatid and chromosomes are aligned in pairs. Each pair of chromosomes has four chromatids and they have a centromere attached in the center holding the four strands together.
2. Nucleolus disassociates and nuclear envelope dissolves.
3. Segments of the closely associated pairs of chromatids may be exchanged with each other (between the pair members) this is called crossing-over. Each chromatid contains the original amount of DNA but now may have "traded" genetic material.
4. The chromosomes separate. Some spindle fibers are forming and some are attaching to the centromeres of the chromosomes. The fibers extend from each pole of the cell.

Metaphase I–

1. In pairs, the chromosomes align at the equator of the cell, with the centromeres and spindle fibers apparent.

2. The two chromatids, from each chromosome, function as a single unit.

Anaphase I–

1. One entire chromosome, consisting of two chromatids, migrates from the equator to a pole. The chromosomes do not separate from each other and retain both chromatids when the reach their pole. At each pole, there will be half the chromosome number. If crossing over occurred in prophase then the chromosomes will consist of original DNA and DNA from a homologous chromosome—now at the opposite pole.
2. The centromere remains intact in each pair of chromatids.

Telophase I–:

1. What occurs in this step, depends on the species involved, as they may revert to interphase or proceed directly to division II.
2. If they revert to interphase, they will only do so partially and the chromosomes will become longer and thinner.
3. nuclear envelopes will not form, but the nucleoli will generally recluster.
4. Telophase is over when the original cell becomes two cells or two nuclei.

Division II—Equational Division—

The chromosome number stays the same, the cells replicate and result in four cells. The events closely resemble the events in mitosis, except that there is no duplication of DNA during the interphase that may or may not occur between the two divisions.

Prophase II—Main features:

Chromosomes of both nuclei become shorter and thicker. The two-stranded nature becomes apparent once again.

Metaphase II—Main features:

1. Chromosomes align their centromeres along the equator.
2. Spindle fibers form and attach to each centromere, extending from one pole to the other.

Anaphase II—Main features:

The centromeres and chromatids of each chromosome separate and begin their migration to the opposite poles.

Telophase II—Main features:

1. The coils of chromatids—now called chromosomes again—relax and the chromosomes become longer and thinner.
2. Nuclear envelopes and nucleoli reform for each group of chromosomes.
3. New cell walls form between the four groups of chromosomes.
4. Each set of chromosomes in the four new cells, has exactly half of the chromosome number of the original number.

ALTERNATION OF GENERATIONS

One member of every original pair of chromosomes end up in each resulting cell, which means each cell has one set of chromosomes. None of the daughter cells is identical to the parent cell. A cell with one set of chromosomes is called haploid and a cell with two sets is called diploid. In sum, a diploid cell undergoes meiosis and results in four haploid cells.

Sex cells, or gametes, of an organism are haploid and when a zygote is formed, the zygote is diploid. This is always true, no matter how many chromosomes an organism might have. In many places, you will find it stated that an organism has n number of chromosomes. In this case, an organism will have 2n chromosomes in its diploid cells and 1n chromosomes in its haploid cells. Alternation of generations refers to a plants life cycle including sexual reproduction that is characterized by alternating between a diploid (2n) sporophyte phase and a haploid (1n) gametophyte phase.

BRYOPHYTES

Bryophytes are essentially nonvascular plants, meaning they do not have xylem or phloem. The habitations of bryophytes are widely varied and include bare rocks in the scorching sun to frozen alpine slopes. Bryophytes include mosses, liverworts and hornworts. These groups of plants require external water, usually in the form of dew or rain. Some bryophytes grow exclusively in dark, damp environments in order to provide moisture. Water is essential for bryophyte reproductive activities. Most mosses have water-conducting cells called hydroids in the centers of their stems, and some even have food conducting cells called leptoids. These cells are not nearly as efficient as xylem and phloem and generally, bryophytes are not very tall plants. The lack of vascular tissue leaves the plant body very soft and pliable. Bryophytes are good nest-making material for birds.

The alternation of generations in bryophytes is quite obvious as the gametophyte generation is the 'leafy' plant that is generally visible. The sporophyte generation that produces spores is located at the tips of the 'leafy' gametophyte generation. The sporophyte generally looks like a slender stalk with a cap on top. While the lifecycles of all bryophytes are similar and even their chromosome number and habituation have similarities; they care divided into three distinct divisions based on the few differences in structure and reproduction.

Division Hepaticophyta—Liverworts

Wort means plant or herb, and in ancient times herbalists thought that some bryophytes—specifically the ones that look like liver lobes—were useful in treating liver ailments. Although, the belief was discounted the name stuck.

Structure—About 20percent of the liverworts have a flattened, somewhat leaf like body called a thallus (plural thalli). The other 80percent, which aren't as common in nature, are 'leafy' and look more like mosses. Mosses are more complex than liverworts. They have one-celled rhizoids on their lower surfaces. The rhizoids look like tiny roots and anchor the plants to surfaces and soil particles. The thalli are the gametophyte generation and develop almost directly from spores. They have smoother upper surfaces and the cell wall corners are thickened.

Thalloid liverworts—*Marchantia* is the best-known species of thalloid liverwort. It can usually be found on damp soil after a fire. *Marchantia* can reproduce asexually and sexually. Asexual reproduction is accomplished through gemmae, which are tiny lens-shaped pieces of tissue that detach from the thallus. Gemmae cups are produced along the upper surface of the gametophyte. Raindrops splash onto the cups and the cups detach and may splash up to 3 feet away from the 'parent' gametophyte. Lunularic acid inhibits the Gemmae from growing, as soon as it is out of the cup, though, the inhibition is removed and each may develop into a new thallus. Sexual reproduction in the *Marchantia* involves the interaction between spores on separate male and female gametophytes. The gametangia are formed on gametophores, or umbrella like structure on long stalks. The stalks are positioned between the grooves of the thallus. The male gametophore is shaped like a disc with a scalloped edge, while the female gametophore looks like the hub and spokes of a wheel. the male gametangia are called antheridia, contain numerous sperms and are produced in rows just beneath the upper surface of the antheridiophore. The female gametangia are flask-like and each one contains a single egg. They are produced in rows and hang with their neck downward from the archegoniophore. Rain will splash and release the flagellated sperm cells. The stalks of the archegoniophores may not be finished growing at the time of fertilization. The zygote, fertilized egg, will develop into a multicellular embryo (a.k.a. immature sporophyte) which is anchored from the tissues of the archegoniophore by a knoblike foot. The foot is connected to the sporophyte (also referred to as the capsule, out of which develops the various type of tissues) by a short, thick stalk called the seta. Inside the capsule, spore mother cells undergo meiosis which results in haploid spores. Some capsule cells do not undergo meiosis, but remain diploid and develop into elaters, which are long and pointy and responsive to changes in humidity. The elaters will twists and untwist rapidly in order to disperse the spores. The young sporophyte will be protected until maturity by a caplike tissue called the calyptra. There are other variations of this cycle in the other thalloid forms. In some floating or amphibious liverworts, the spores are freed only as the thallus decays.

"Leafy" liverworts—These liverworts are usually found in tropical jungles of fog belts. They have two rows of partially overlapping 'leaves' in which the cells contain distinctive oil bodies. They are of course, not true leaves. They do have folds and lobes that collect water and usually house tiny animals. The male and female gametangia produce antheridia and archegonia in cuplike structures. When the sporophytes mature they will germinate and produce a protonema which has photosynthetic cells and will develop into a new gametophyte plant.

Division Anthocertophyta

Hornworts—The mature sporophytes of a hornwort look like miniature green cattle horns. Their gametophyte generation looks similar to filmy versions of thalloid liverworts. Hornworts are rare in artic regions and are usually found in moist, shady areas, although some do grow on trees. They have between one and eight chloroplasts—usually just one—and the chloroplasts have pyrenoids similar to green algae.

Asexual reproduction—They can reproduce asexually by fragmentation or by developing lobes that are separate from the main part of the thallus.

Sexual reproduction—Hornworts also reproduce sexually and the plants can be unisexual, like mosses and liverworts, or they can be bisexual, meaning the archegonia and antheridia are on the same plant. The sporophytes are distinctive in form and contain numerous stomata. They do not have stalks, setae, and look instead like tiny broom handles rising out of the gametophyte generation. Meristematic tissue at the base of the sporophyte continually increases the length of the sporophyte until conditions are favourable and the axis (central core) undergoes meiosis to produce spores. The sporophyte tip will split and the spores will be dispersed as the horn peels into ribbon like segments.

Division Bryophyta–Mosses

Subclasses–There are three subclasses of mosses: peat mosses, true mosses and rock mosses. They share similar reproductive and life cycles and they are all distinct from other plant organisms. Moss 'leaves' do not have mesophyll, stomata or veins and all the 'leaf' cells are haploid. The cells in the 'leaf' do contain numerous lens-shaped chloroplasts, except in the midrib. Mosses have transparent water-storage cells that chloroplasts, yet aid in absorbing water and thus providing moisture to the avascular plant. Mosses also have root like rhizoids which do provide for some water absorption. However, most of the water for the plant travels up the moss surface by way of capillarity.

Asexual reproduction–This reproductive process does not rely on a sexual cycle and its alternation of generations, instead it has been demonstrated that moss fragments can produce protonemata which will 'bud' and develop into gametophyte mosses.

Sexual reproduction–Mosses produce gametangia on the same plant or on separate plants. Commonly, the gametangia are produced on the same plant. The multicellular antheridia and archegonia are produced at the tips of 'leafy' shoots. The individual archegonium has a cavity, the venter with a single egg, and a neck through which the sperm gains access to the egg. Sperm cells are produced in the antheridia. Upon fertilization, the zygote develops into an embryo that remains attached to the gametophyte by an embedded foot. The embryo develops into a sporophyte with a capsule and a seta, a stalk. The gametophyte produces a calyptra which partially covers the capsule. Inside the capsule, spore mother cells undergo meiosis and produce spores which are then released through the teeth of the peristome located at the tip of the capsule. Until spore maturity, the peristome is protected by an operculum, which will fall off at maturity. After the spores germinate, protonemata will 'bud' and develop into gametophyte mosses.

VASCULAR PLANTS

Ferns and Relatives

These plants are seedless plants, but unlike the bryophytes, they do have vascular tissue (xylem and phloem). Because of the presence of vascular tissue, the leaves of ferns are their relatives are better organized than the mosses and liverworts.

Four divisions:

Division Psilotophyta

Overview:

The sporophytes in this division have neither true leaves nor roots, their stems and rhizomes fork evenly.

Whisk ferns—Whisk ferns are the simplest of the vascular plants. They consist of evenly forking stems with small protuberances called enations. They lack leaves and roots. These ferns have a central vascular cylinder composed of xylem and phloem.

Reproduction—Whisk ferns reproduce via gametangia. Spores will germinate into tiny saprophytic gametophytes on the surfaces which antheridia and archegonia are scattered. The resulting zygote will develop a foot and a rhizome. An upright stem is produced when the foot separates from the rhizome.

Division Lycophyta

These plants have stems that are covered with photosynthetic microphylls. Microphylls are leaves with a single vein and a trace that is not associated with a leaf gap.

Club mosses –there are two types of club mosses that still have living representatives. Ground pines will develop sporangia in the axils of sporophylls. Each gametophyte may be capable of producing several sporophytes. The second type of club moss include the spike mosses. These plants are heterosporous and have a ligule, or little tongue, on each microphyll. The microspores will develop into male gametophytes with antheridia (sperm producing); while the megaspores will develop into female gametophytes with archegonia (egg producing). The biggest difference between the ground pines and the spike mosses is the presence of the ligule (spike mosses) and the spike mosses produce two types of spores and gametophytes (heterospory).

Quillworts

Quillworts are found partially submerged in water for at least part of the year. Their microphylls (leaves) look somewhat like porcupine quills although they lack the rigidity of an actual porcupine quills. The microphylls arise from a corm like base. The corm base has a cambium that will remain active for many years.

Club moss spores have many uses including flash powder, medicine, talcum powder, as well as ornamental uses and novelty items.

Division Sphenophyta

The plants in this division have ribbed stems that contain silica deposits in the epidermal cells. Their scale-like microphylls lack chlorophyll. However, the silica in the stems makes these plants useful for scouring. Horsetails and scouring rushes occur in both the branched and unbranched forms. Either way, they are jointed stems with small whorls of scale-like leaves at the base of the plant.

Horsetails, or Equisetum, are hollow in the center of their stem which contains cylinders of carinal and vallecular canals. The hollow stems arise from extensively branching rhizomes just beneath the surface of the soil. Some types of horsetails have non-photosynthetic stems. They reproduce via strobili, cones, that are produced in the spring in all species. Horsetail spores have ribbon like elaters that are sensitive to humidity. An interesting note about horsetails; usually equal numbers of male and female gametophytes are produced by these plants, however, the female gametophytes may become bisexual and the development of more that one sporophyte from a gametophyte is common.

As mentioned, horsetails have been used for scouring. They can be eaten if the silica is removed, however, they really don't make for a feast. Plants from this division have also been used for medicinal purposes including as a diuretic, treatment for tuberculosis and treating venereal diseases. Horsetails have been used as an ingredient in shampoos and metal polish.

Division Pterophyta

Ferns are the most common and best recognized examples of the vascular seedless plants. The leaves of ferns have megaphylls—or leaves with more than one vein and a leaf trace/leaf gap association—which are large and usually subdivided into many lobes. Hence, the 'finger' look to a fern plant. Fern fronds, as the leaves are called, typically are dissected and feathery in appearance. However, realize that they do vary quite a bit as far as external structure and form goes. Fern fronds start as croziers or fiddleheads which unfurl into the main leaf form. On the underside of the frond, patches of sporangia can be found. These sporangia are usually in sori, which are clusters; and are sometimes covered by a clear flap called the indusium. This setup allows for the maximum protection of the sporangia, but allows for distribution. Individual sporangia have an 'spring-loaded' annulus which allows for mature spores to be catapulted out of the sporangium. The gametophytes of ferns are called prothalli and develop after the spores germinate. Prothalli contain both archegonia and antheridia and only one zygote develops into a sporophyte.

Ferns are used for ornamental purposes as well as stuffing materials for bedding in tropical regions. Fern fronds are used in weaving baskets and hats, brewing some types of ale and numerous folk medicine practices.

SEED PLANTS

There are two main subdivisions of seed plants—the ones without covered seeds, the gymnosperms, and the ones with covered seeds, the angiosperms.

GYMNOSPERMS

Gymnosperms are plants that do not flower and do not bear their seeds in an enclosure such as a fruit. The seeds are produced on the surface of the sporophylls or similar structures until they are dispersed. The sporophylls are usually arranged in a spiral on the female strobili (cones) which develop at the same time as the smaller male strobili. The male strobili produce the pollen which will fertilize the ovules in the female cones. The ovule contains a nutritious nucellus which is itself enclosed in several layers of integument. The integument layers will eventually become the seed coat, after fertilization and further development of the embryo takes place.

Gymnosperms are classified into one division and three subdivisions: division *Pinophyta* with subdivision *Cycadicae* which includes the palm like cycads; subdivision *Pinicae* which includes conifers and class *Ginkgoatae* the Ginko trees; and subdivision *Gneticae* which includes the gnetophytes.

Subdivision *Pinicae*: Conifers—pine trees and evergreens to the layman.

Structure and form—For the sake of discussion we will look at Pines, which are the largest genus of conifers. Pine needles are their leaf structures. They are usually arranged in clusters or bundles of two to five leaves (needles), although some species have as few as one or as many as eight leaves in a cluster, the clusters are sometimes referred to as fascicles. Each needle is covered with a thick cuticle over the epidermal layer and a layer of thick-walled cells just beneath the epidermis called the hypodermis. The stomata on the epidermal surface are sunken and are surrounded by an endodermis. The mesophyll cells do not have the wide air spaces as broadleaf and flowering plant leaves. Resin, and resin canals develop noticeably throughout the mesophyll cells.

The canals are tubes in which resin is secreted. Resin is both aromatic and antiseptic and helps to prevent fungal infections and deter insect attacks. Some conifers produce resin in response to injury. The fascicles, needle clusters, will fall off every two to five years after maturing. They do not, however, fall off all at once and unless diseased, will not look bare like other flowering trees. The secondary xylem, wood, in conifers varies in hardness. Most gymnosperm wood consists of tracheids and has no vessel members or fibers as do flowering trees. Therefore the wood lacks thick walled cells. Conifer wood is considered to be softwood, while the wood of broadleaf trees is considered to be hardwood. The xylem rings in conifers are often fairly wide as a result of rapid growth. Both vertical and horizontal resin canals can be found throughout the wood. Pine phloem lacks companion cells, but has albuminous cells that perform similar function for the phloem. The roots of pine trees are always found in association with mycorrhizal fungi. The fungi perform functions for the roots, which enable normal growth. Pine trees can be found in all types of environments and ones of opposite extremes.

Reproduction—There are two kinds of spores produced by pine trees. The microspores are produced in the smaller male strobili which develop at the tips of lower branches. At the base of the male cone, the microsporangia develop in pairs and give rise to four-celled pollen grains. Millions of pollen grains are produced per cone. The female cones are larger and produce the megaspores. They are formed in ovules at the bases of female cone scales. A pore, called the micropyle, allows for access by the sperm (pollen grain) to the ovule. The pore is formed by the overlapped layers of integument protecting the ovule. Each megaspore develops into a female gametophyte. The archegonia is contained in the mature female gametophyte. Prior to the maturity of the archegonia, pollen grains will become lodged in the cone scales in sticky pollination drops. The pollen grains develop a long pollen tube that digests its way down to the developing archegonia. Upon arrival at the archegonia, two of the original four cells in the pollen grain will migrate into the tube. One sperm will unite with the egg cell to form a zygote. The zygote will continue developing and will become a seed embryo with a membranous wing formed from part of the cone scale. The seed embryo is ready for distribution and upon landing will germinate and become a new tree.

Class *Ginkoatae:* Ginkgo trees have small fan-shaped leaves with veins that evenly fork. They have similar reproductive cycles to that of the conifers with the exception that the edible seeds are encased in a fleshy covering. The covering smells like rancid butter at seed maturity.

Subdivision Cycadicae:

Cycads—These plants look like little palm trees with unbranched trunks and large crowns of pinnately divided leaves. Their strobili and cones are quite similar to those of conifers, however, their sperms have numerous flagella—much unlike conifers.

Subdivision Gneticae

Gnetophytes—These plants have vessels in their xylem. Most of the species are in the genus Ephedra and have jointed stems and leaves that are nothing more than scales. Sometimes the plants in this genus are called joint firs, as they look like jointed sticks. The plants in this subdivision are adapted to unusually dry environments. They produce tiny leaves in groups of twos and threes, which turn brown as soon as they appear. Male and female strobili may occur on the same plant.

RELEVANCE TO HUMANS

Conifers are sources for paper products and lumber materials. The resin from conifers has historically been used as sealing pitch, turpentine, floor waxes, printer's ink, perfumes, menthol manufacture and rosin for musical instruments. Ginko leaves are used medicinally as are plants from the genus Ephedra. Arrowroot starch was once purified from a cycad species. Teas have been made from conifers.

Angiosperms—Flowering Seed Plants (Covered Seed Plants)

Angiosperms are plants that have seeds encased in a protective covering. That covering is the ovary which is part of the flower structure and distinguishes angiosperms from gymnosperms, the other seed plants. So it can be said that angiosperms are also flowering plants. There is one division of angiosperms, Magnoliophyta, which is divided into two classes: monocots and dicots. Angiosperms, like gymnosperms, are heterosporus, which means they produce two types of spores and their sporophytes are more dominant than those of gymnosperms. At maturity, the female gametophytes are reduced to a few cells and are completely enclosed within sporophyte tissue; while the male gametophytes consist of a binucleate cell with a tube nucleus which forms a pollen tube much like the one formed in gymnosperm pollination.

DEVELOPMENT OF GAMETOPHYTES

Angiosperm gametophytes develop in separate structures, sometimes on the same plant. The embryo sac, or female gametophyte, develops in the ovule which is surrounded by integuments. The integuments will later become the seed coat after fertilization has occurred. The male gametophytes, pollen grains, develop in the anthers.

Pollination—The process of transferring pollen grains from the anther to a stigma is called pollination. This may occur via wind, insects, birds or other agents. Flowers may be geared toward certain pollinators. Flowers pollinated by bees are usually sweet and fragrant. Their colours are usually blue and yellow. Beetles are attracted to flowers with strong odors and dull colours or white flowers. There are fly-pollinated flowers that smell like rotting meat. Moths are drawn to pale yellow or white flowers. Birds, akin to bee preferences, like sweet nectar and fragrant flowers with bright colours. Orchids have unusual pollination mechanisms, and in some they literally grab an insect and manually attach two pollen packets to the hind end of the insect before releasing it.

Fertilization and seed development—The pollen grain contains two sperm nuclei (formed from a generative nucleus that splits) and one tube nucleus. The tube nucleus forms a pollen tube that grows through the stigma, the style and into the ovary via the micropyle. Upon arrival at the ovary, one sperm nuclei will fertilize the egg and form a zygote. The other sperm nuclei will fuse with the polar nuclei of the egg in order to form a 3n endosperm nucleus. This will be conserved as food for the plant embryo or may become part of the seed. Some species will end up with 5n, 9n, or 15n endosperm tissue, depending on how the embryo sac develops.

Parthenocarpy—Some fruits can develop without the development or fusion of gametes, but with the otherwise standard structures being involved. These fruits are called parthenocarpic fruits. Not all 'seedless' fruits are parthenocarpic.

CLASSIFICATION TRENDS IN FLOWERING PLANTS

Specializations in flowering plants include an overall efficiency in plant growth, by reducing the number of parts; fusion of parts; the appearance of compound pistils that are composed of several carpels; various positioning of the ovary with respect to the receptacle including inferior and superior ovaries; irregular flowers and unisexual flowers. A monoecious species of flower has both male and female flowers on the same plant, while a dioecious flower would have separate plants for male and female flowers. All of these are characteristics that can be looked at in the classification and categorization of flowering plants.

RELEVANCE TO HUMANS

Beyond ornamental uses, flowering plants constitute much of what we eat, parts of the clothes we wear, the wood in our homes and furniture and the medicines we consume. Flowering plants are everywhere and thus have a million uses. All fruit comes from flowering plants, obviously, and think of how many just in the edible category there is, not to mention all of those that aren't for eating. Stop and think for a minute on the plants that you encounter in daily life, chances are good they came from a flowering plant.

Chapter 7

Regulatory and Ethical Aspects

GENE THERAPY

A "perfect" cure of a genetic disease would be to provide the patient's cells with a corrected copy of the faulty gene. The first examples of success with such "gene therapy" now are available. Gene therapy is the correction of a heritable disease by the addition of a functional gene. Technically, the term can be divided into two categories: "germ-line gene therapy" and "somatic cell gene therapy". On ethical grounds, the latter is the only kind of gene therapy being considered in humans and involves the treatment of different cells in the body, but does not allow inheritance of the genetic changes; it is analogous to an organ transplant. For a disease to be a candidate for gene therapy the biochemistry of the disorder must be well understood, a cloned functional gene must be available, and there must be means available to transfer that gene into a tissue where the gene will be beneficial. Current conventional treatment protocols must be nonexistent or inadequate. Somatic gene therapy entails isolation of cells from the patient, in vitro transformation of these cells to introduce the desired gene, and reintroduction of transformed cells into the patient.

A virtual requirement for successful gene therapy is a self-propagation of the transformed cells in the patient, as only then the product of the restored gene will continue to be produced and will continue to circulate in the body (remember that all proteins have a limited lifetime). A suitable type of cells for introduction of a functional gene are bone marrow stem cells, which divide and can differentiate to various types of white blood cells (lymphocytes). An early example of gene therapy applied to humans was the case of a young girl, who was born with adenosine deaminase (ADA) deficiency. This deficiency leads to an impairment of production and maintenance of two types of lymphocytes, yielding a defective immune system. A functional ADA gene was introduced in T cells (also a type of lymphocytes), and the transformed T cells were introduced into the patient. She has responded well to the treatment, and apparently is leading a normal life. Subsequent young patients have fared equally well as a result of the gene therapy treatment. Encouraged by this success, gene therapy is now applied (mostly on an experimental basis) to treat cystic fibrosis. In this case, the functional gene is delivered to lung tissue by a viral vector that has been disabled so as to not be infectious.

Other potentially suitable cell types for gene therapy are skin fibroblasts (which grow well in culture), and hepatocytes (liver cells). The problem with both cell types is that they are restricted to a certain part of the body, and do not circulate. Therefore, gene therapy using such cells will be most effective if it is to treat diseases that affect mostly the skin and the liver, respectively, or that are metabolic in nature and affect the level of certain compounds in the bloodstream.

One problem with gene therapy is that one does not have control over where the gene will be inserted into the genome. The location of a gene in the genome is of importance for the degree of expression of the gene and for the regulation of the gene (the so-called "position effect"), and thus the gene regulatory aspects are always uncertain after gene therapy. Another problem is that gene therapy is still experimental and is not without risks. One example was the 1999 death of Jesse Gelsinger, a gene therapy patient who lacked ornithine transcarbamylase activity.

The vector by which the appropriate gene will be introduced into the body is of major importance for the success of gene therapy. The earliest vectors were based on the murine leukemia virus, and carried their genetic passengers into dividing cells. The drawback of this is that most of the body's cells are non-dividing. The next generation made use of adenoviruses, the same kind of viruses that cause the common cold. These viruses had a higher rate of delivery, but the immune system quickly kicked the foreign material out of the body. Lentiviruses, including HIV, promise to incorporate their passenger genes into non-dividing cells, but the use of an attenuated strain that is closely related to deadly siblings causes much concern. Probably the most promising group of viruses in terms of gene therapy are the adeno-associated viruses (AAVs) that cause no known disease in humans; however, they are difficult to produce in mass quantities.

A matter of mostly ethical consequence is the question whether human gene therapy may involve cells that will be transmitted to the next generation (egg or sperm cells). In the case of transformation of bone marrow cells, the patient will be cured, but he or she will still be a carrier of the disease in the sense that progeny will inherit non-transformed cells. This problem would be avoided if generative cells would be transformed as well. However, there is not yet a consensus on whether this would be ethically correct. Even though most people may agree that transformation of generative cells to combat life-threatening diseases should be allowed, it is difficult to decide on where to draw the line.

After ethical issues would be resolved, a practical consideration would be how one should go about altering germ cells in humans. This does not appear to be simple. In mice three methods have been described to work reasonably well:

1. Manually inject 100-1000 copies of a gene into the pronucleus of a recently fertilized egg. The embryos are then transferred back to a mother mouse, and usually a few percent of the offspring will inherit the gene and pass it on to succeeding generations. Of course, even though this works well for mice, its low

percentage would not make it suitable for humans. Also, it can be carried out only at the single-cell stage of the fertilized egg, making it useless for gene therapy after birth.

2. Transform embryonal stem cells. These cells are early embryonic cells that can grow in culture. When mixed with non-transformed cells of the growing embryo, they are capable of giving rise to all cell types of an organism, including germ cells. However, it is unpredictable to what cells the embryonic stem cells will give rise in any particular case, and again this requires treatment at the embryo stage. Thus, this also is not particularly applicable to humans.
3. Genes can be introduced into retroviruses, which have a certain probability to integrate into genomes, taking inserted genes with them. But again, this is unpredictable in its frequency and target. Thus, for better or worse, human gene therapy of germ cells still may be a long way off.

The general success of gene therapy experiments, however preliminary and limited in scope, now has led gene therapy more into the mainstream of medical science. The federal government has moved towards relaxing its scrutiny of human gene therapy experiments, so that the approval process of most gene-therapy protocols are much quicker and simpler. Not all gene-therapy protocols need to be subjected to public examination before the Recombinant DNA Advisory Committee (RAC) anymore. Only if new and cutting-edge experiments are proposed, or if important safety or policy issues are concerned, researchers may be asked to appear before the RAC. The reason for the change in regulation mostly is caused by the fact that many of the protocols are rather standard, and there is no need to go through a lengthy approval process for something that has been approved in essentially identical form already.

However, several unsuccessful gene therapy treatments in the past year have caused public concern, and a tightening of the protocols and improvement of patient monitoring is being considered.

ETHICAL CONSIDERATIONS

Medical applications of biotechnology may have far-reaching ethical consequences. For example, in 1993 an announcement was made that "scientists had cloned human embryos"; three copies of human embryos were created outside the body, and were allowed to develop for six days. A more recent stir was the 2001 announcement by Advanced Cell Technology that cloned human embryos of 4-6 cells had been grown. In essence, human cloning qualitatively is not new, as identical twins are clones. However, the test-tube approach is different in that embryos are selected (which is not the case for identical twins) and can be divided many times, to create a large number of embryos with genetically identical makeup. Since then, developments have been rapid, and there no longer are technological reasons why one could not clone a human. A human embryo reportedly has been cloned in Britain in 2004.

Bioethics also extend into the traditional medical arena. With the development of life-extending tools (including mechanical ventilators, exotic drugs, organ transplants, and artificial nutrition and hydration devices), one must ask the question whether we should keep a person alive just because we can. And who should answer that question? Also, should particular medical products (such as human growth hormone) be available to everyone, or what are the grounds for selection? Or who should have access to anyone's DNA fingerprint? And how ethical is the business standpoint of pharmaceutical companies, who invest a lot of money to develop "luxury" medicine and treatment for the rich in the Western world, but who have invested 100-fold less in new drugs for malaria, cholera, and other lethal maladies of the tropics? These are issues that clearly go beyond what is covered in the classroom, but important to think about and form an informed opinion about.

STEM CELLS

Although mouse embryonic stem (ES) cells have been isolated more than 20 years ago and have successfully been used in creating transgenic mice, it was not until 1998 when Dr. James Thomson from the University of Wisconsin, Madison succeeded in isolating the first human ES cell line. Embryonic stem cells are cells derived from the inner mass of the blastocyst (4-5 day old human embryo) that are capable of self-reproduction for long periods of time and that can yield specialized cells that constitute various tissues and organs.

Because of the pluripotent nature of stem cells, researchers and patients alike put great hopes into the use of ES cells for treating diseases such as Parkinson's, Alzheimer's, diabetes and repairing damaged hearts and spinal cords. As often is the case with exciting new technologies, ES cells are surrounded by a cloud of political, ethical and other issues that make them very controversial.

THE RELEASE OF GENETICALLY ENGINEERED EUKARYOTES INTO THE ENVIRONMENT

The "early" applications of genetic engineering all involved expression of human genes (such as the insulin, growth hormone and interferon genes) in bacteria that were kept in the laboratory. The level of expression was up to a million protein molecules per bacterial cell, which means that about a mg of protein could be extracted from 100 ml of cell culture. This was a marked improvement over the traditional method of extraction of relatively rare enzymes from tissue. For many purposes, it suffices to grow genetically engineered organisms in the laboratory, and to obtain the desired product from them. However, laboratory containment of production-scale numbers of transgenic plants and animals is impractical, and such genetically altered species will need to be released. Therefore, it is important to consider the requirements that need to be met for a genetically modified organism to be released in the environment.

The first genetically engineered organism that was approved for release in the US in the early nineties was a bacterium, *Pseudomonas syringii,* which, in contrast to most of its naturally occurring kin, does not make a protein that acts as a nucleation site for ice formation during light freezing conditions. Thus, if the genetically engineered bacteria rather than the naturally occurring ones are sitting on (and in) a plant leaf, ice formation (leading to rupture of cell walls and to plant damage) will not easily occur. Natural *Pseudomonas syringii* strains with properties identical to the genetically modified one do occur, so one can argue that no new bacterial characteristics are added to the ecosystem.

Around the same time, field tests were done with genetically modified plants. In 1992, transgenic tomato that expresses the coat protein from a plant virus, the tobacco mosaic virus, was tested. Plants that express this coat protein (obtained by transforming tomato protoplasts with the virus coat protein gene, and regenerating the protoplast to an entire plant) are less sensitive to the tomato mosaic virus and the tobacco mosaic virus. The insertion and expression of the coat protein gene in tomato does not result in any loss of productivity of the plant, whereas virus infection of non-resistant plants can decrease yields by 20percent or more. Thus, the insertion of the gene for the tobacco mosaic virus coat protein into tomato leads to an increased resistance against two viruses without affecting the yield of the crop. The reason why viral coat protein expression in the plant leads to increased resistance of the plant against the "real" virus is not yet understood.

As will become clear in the next few paragraphs, the rate of release of genetically engineered organisms during the past decade has been mindboggling. However, first we need to consider what may be the effects when one introduces genetically engineered organisms into the environment.

1. Introduction of a gene whose product metabolizes the herbicide,

2. Introduction of a highly expressed copy of the gene for the receptor protein (thus "catching away" all herbicide molecules, and still have enough receptor protein without herbicide to function with), or
3. Introduction of a gene that contains a mutation in the receptor protein so that the receptor protein has a decreased affinity for the herbicide without affecting the functional activity of the protein.

In the United States, herbicide-tolerant soybeans became available to farmers for the first time in 1996. This has proven to be quite a success story. Within two years, over 4 million acres (40 percent of the U.S. soybean acreage) had been planted with herbicide-tolerant soybeans — making soybeans the number one bioengineered crop in the United States. Why have U.S. farmers taken to the herbicide-resistant soybeans and similar crops? It is because they have found that using these bioengineered seeds reduces the need to plow their fields to control weeds; decreases the amount of chemical herbicide they need to use; produces higher crop yields; and can deliver a cleaner and higher quality harvest. Total herbicide sales have not increased, indicating that environmental impacts are limited. Pesticide tolerance has been introduced primarily by means of introducing the gene for Bt toxin into plants. Corn provided with genetic protection from the European corn borer by insertion of the appropriate Bt gene was approved in the United States in August 1995. Fields of corn with Bt protection, on average, have 7-11 percent increase in yield per acre in comparison to more conventional corn, and the amount of sprayed pesticide has decreased, thus decreasing the environmental impact of agriculture.

Recombinant organisms to be "released" also include farm animals. Transgenic pigs, sheep, and cows have been produced, mostly by injection of DNA into embryos of the animal. This appears a relatively efficient procedure: even in early stages of the technology, of 92 lambs that were born from injected embryos, six were found to be transgenic. The genes transferred in this particular experiment were those for a human blood clotting factor, and for a protease inhibitor. These proteins can be excreted in the milk of the animals, even though it should be kept in mind that these proteins generally will not reach the bloodstream of the individual drinking the milk: proteins are broken down upon digestion before uptake into the body. Therefore, as indicated before, the desired protein has to be isolated from other components in the milk, and injected into patients.

Release of transgenic crops and animals in some respects is similar to the release of a "new" organism into the environment. The effects may be difficult to predict: there are success stories, but (often unintentional) introduction of new species into an ecosystem in some cases has led to huge disasters. A fungus introduced into America from Asia killed almost all of North America's chestnut trees. Another fungus has eliminated most Dutch elm trees from the Eastern US. The myxomatosis virus introduced in Australia (by Australian scientists) almost completely annihilated that continent's rabbit population, where this species had become a major pest within a century after its introduction in Australia. However, it should be mentioned that the rabbit strikes back, and has become essentially resistant to the virus; another virus has now been "accidentally" introduced in Australia. More than half of the insect pests in the US today come from abroad. Similarly, starlings, house sparrows and gypsy moths are all introduced animals that America could have lived without. By the same token, most of the USA's major crops, including soybean, wheat and also rice, are not indigenous to America. Such arguments are valuable reminders of biotechnology's potential to do great good or great harm, and thus one should examine very carefully the potential effects of the release of genetically engineered organisms into the environment. On the other hand, it should be realized that in many cases hybrids with properties close to those of certain genetically altered organisms could be selected by repeated crosses, as has been done virtually throughout history. Thus, it makes no sense to over-react when encountering an organism that is genetically engineered; factors such as the nature of the introduced gene (and its product) need to be evaluated first before any judgment on its potential environmental danger or on its merits can be made.

One particular issue pertaining to the safety of release of genetically modified organisms is whether desired characteristics in crops can confer adaptive advantages to weedy species. Spreading of plant genetic material is very easy. Bees are important pollen vectors over a range of distances and farm-to-farm spread of plants such as canola (rapeseed) that have closely related wild relatives. Pollen can also travel for miles in the wind. If crossing with wild plants and native species, herbicide or pesticide resistant weeds might result. Indeed, in Canada canola-related weeds have been found that have become resistant to three types of herbicides. It is not only related plants that may receive the transgene. There is also a report of gene transfer from genetically engineered rapeseed to bacteria and fungi in the gut of honey bees. Therefore, it is clear that one cannot guarantee the absence of spreading of transgenes that have been introduced. The question that then needs to be addressed in the evaluation of the risks of release of a genetically modified organism is whether such a spread would pose unacceptable environmental risks.

Apart from a rational analysis of potential risks, public opinion is also an important factor in the fate and success of transgenic plants and animals. After a decade of research, development, testing, and approvals, Calgene's Flavr Savr genetically engineered tomato hit the market place in the nineties. This tomato is slow-ripening due to decreased ethylene production, and thus it can be picked later and still be sold in good shape in the grocery store. This was the first genetically engineered food product to be marketed. However, there was a concern about the public perception of the product, and the Flavr Savr tomato no longer is marketed commercially. Another example of where public perception had a major effect was the case of Starlink corn: In 2000, in tacos traces of a Bt corn variety (DNA fingerprinting in action!) were found that had not yet been approved for human consumption (pending the outcome of allergen tests) but had been approved for use as animal feed. A public outcry followed. Even though Starlink corn was approved for public consumption soon thereafter and no proven cases of allergies due to Starlink have been found, this was poor public relations for the biotechnology industry. Currently, no corn varieties will be approved for animal feed if they have not been approved for human consumption as well.

The challenge for agrobiotech companies has been to find profitable products that are safe, that provide benefits to the consumer, and that do not elicit negative responses from the popular press. The bottom line is that public acceptance of genetically engineered food products that are consumed without further processing (such as tomatoes) is relatively poor, but no one seems to be very concerned about genetically modified plants (such as soybeans) that do not serve as primary food source but whose products are in most foods. The time to be invested in development and testing is 4-10 years, or even longer. However, with more experience with the approval process and marketing, this time may get shorter in the not-so-distant future. Nonetheless, do not look for any special indications on genetically engineered products: for better or for worse, there is no requirement that genetically engineered food products are labeled as such for the consumer.

Guidelines have been prepared by both national and international institutions for screening and characterization procedures of recombinant DNA organisms. The most well-known are the 1976 NIH (National Institute of Health) guidelines for research involving recombinant DNA molecules, formulated after a meeting of scientists at Asilomar. These guidelines were very strict, since at that time there was not much experience with the potential hazards of biotechnology, and the guidelines were (rightfully) targeted at a "worst-case scenario". After a few years it became clear that many of the risks were initially overestimated, and the guidelines relaxed over the years. Today some 90percent of the experiments involving recombinant DNA are exempt from the guidelines. However, now new regulatory concerns have emerged: in some cases the approval for various aspects of release of genetically engineered organisms or the production of drugs etc. by transgenic organisms is spread out over several federal agencies (Food and Drug Administration, the Environmental Protection Agency, and the US Department of Agriculture), so that product approval is a long and tedious process. Although obviously the streamlining of the approval process must not result in a loss of thoroughness and quality of this process, it is obvious that one single agency can do a better and cleaner job than several agencies that often work in parallel (or antiparallel) with respect to each other. In addition, the workload of some agencies has increased dramatically over the last few years (now biotechnology is coming of age), while the staffing of the agencies has not kept pace.

In an attempt to streamline the approval process for field testing of genetically engineered plants, the USDA no longer requires submission and approval of a description of the method for conducting the field tests before the tests actually take place. Now for routine tests, one only needs to notify APHIS (Animal and Plant Health Inspection Service, a branch of the USDA) as long as the field tests follow basic guidelines. Waiting for approval is not necessary in these cases. To qualify for this "express-lane treatment", the researcher must certify that

1. The transgenic plant is one of the following species: corn, cotton, potato, tomato, soybean, or tobacco;
2. The transferred gene is stable;
3. The process will not produce disease in the transgenic plant;
4. The process introduces no infectious material to the plant;
5. The process poses no significant risk of creating new plant viruses; and
6. The transgenic plant does not contain any functionally intact genes from human or animal pathogens.

This new notification procedure reduces approval time, cuts cost, encourages biotechnology innovations, and focuses USDA resources on the areas of greatest complexity. In another move to limit unnecessary paperwork, USDA allows specific transgenic plants to be removed from regulation after adequate field testing has been completed.

RELEASE OF GENETICALLY ENGINEERED MICROBES

Thus far, release of genetically engineered microorganisms has not been looked upon favourably by scientists and regulatory agencies. Microbes cannot be easily tracked in the environment, and it is felt that risks associated with release of genetically altered microbes are too large. Instead, naturally occurring organisms may be fished out of their natural habitat, enriched in the laboratory, and released at a different location. This may be done, in the case of in situ bioremediation.

Many transgenic microbes are kept in laboratories, and in many instances, the transgenic organisms are bacteria that are grown in large batches in the laboratory and will not get into the environment. However, what about the potential for laboratory accidents and the chance of "escapes" of genetically engineered bacteria? Depending on the known or potential pathogenicity of the host organism or of the transgenic organism, various levels of containment are generally used. In many cases, laboratory strains of *E. coli* are used, which have no chance of survival outside the laboratory, and will also no longer be able to propagate in the human's intestines, where they originally were isolated from. This selective loss of properties is based on the inactivation of a number of different genes, and it can be excluded that mutations can occur within a normal time frame to bring the lab strains back to their wild-type phenotype. However, in various cases a high level of containment of the microorganism is necessary.

When discussing the risks associated with biotechnology and the release of genetically engineered organisms, one has to keep in mind that there are risks involved with almost anything. We ride our bicycles or drive a car, use energy from a nearby nuclear power plant, live in a flood plain, get a suntan, or maybe we even smoke, and still may be worried more about genetically engineered foods than about any of the daily hazards we have chosen to take for granted. Our choices may be irrational because we may make decisions based on emotions rather than facts. Some of the factors triggering our emotions may have to do with whether the risks are voluntary (more acceptable) or involuntary (less acceptable), whether we are in control or not, whether the risks are familiar or not, and whether the risk is natural or man-made. However, this may be pretty misleading. For those of you who like(d) cabbage, mushrooms, and peanut butter: did you know that at least for bacteria the carcinogenic potential of these presumably healthy items is very high? Often the media are keen on just reporting "facts" that will draw the attention of the reader, but there is not necessarily a critical comparison with other items.

Risks and benefits need to be put together and compared. Are we prepared to take certain risks to enjoy particular benefits? Depends on the risks, and depends on the benefits, you may answer. The same may be true with genetically engineered materials. What are we comfortable accepting as risk, and what benefits do we expect from it? This may be an important question to consider it clearly may have a different answer for any of us individually, but we must think about it rationally.

TRANSGENIC ANIMALS

HUMAN PROTEINS

In some cases overexpression of human genes in bacteria does not yield a protein that is functionally active in humans. The reason for this is that some proteins need to be post-translationally modified before they are active. Bacteria generally lack the specific enzymes recognizing the human protein sequences that need to be modified, and thus the bacterially produced gene product will differ from the native one. To counter this problem, certain human genes can be introduced into farm animals, and when these genes are expressed in the mammary glands of the animals, the post-translationally modified protein can be isolated from milk, tested whether its post-translationally modified product is identical or at least very similar to the native human one, and if so, be developed as a pharmaceutical. The genes for two different human blood clotting factors have been hooked up to sheep and pig regulatory sequences that causes expression in mammary tissue; after transformation of sheep or pig embryos, genetically engineered animals have been selected that produce milk with a large percentage of human blood-clotting factor.

This protein can be isolated from the milk, purified, and marketed. Similarly, transgenic rabbits have been created that produce human interleukin-2, which is a protein stimulating the proliferation of T-lymphocytes; the latter play an important role in fighting selected cancers.

Other human proteins that have been expressed in transgenic animals include: anti-thrombin III (to treat intravascular coagulation), collagen (to treat burns and bone fractures), fibrinogen (used for burns and after surgery), human fertility hormones, human hemoglobin, human serum albumin (for surgery, trauma, and burns), lactoferrin (found in mother milk), tissue plasminogen activator, and particular monoclonal antibodies (including one that is effective against a particular colon cancer). Animals mostly used for this work are pigs, cows, sheep, and goats.

The amounts of milk needed to provide a national supply of these pharmaceuticals are really very reasonable. Assuming the animals produce 1 g of the protein per liter milk and one has a purification efficiency of 30percent (that is, 30percent of the protein is recovered in the pure sample), then a pig can produce 75 g of protein per year, a goat 100 g, a sheep 125 g, and a cow 3 kg. As the national need of blood-clotting factor IX is 2 kg / yr, one cow per country can do the job. For other proteins the demand is larger, but nonetheless a limited number of animals is all one now needs to meet the national demand for pharmaceutical proteins that used to be astronomically expensive.

DOLLY AND POLLY

A fairly large stir was caused in the popular media when a group in Scotland associated with the Roslin Institute and with PPL Therapeutics announced in early 1997 that a lamb, Dolly, had been born that had been cloned from a single cell taken from her mother's udder. Of course, cell division of vegetative cells to eventually yield a new organism with the same genetic makeup as the parent is pretty usual among "lower" organisms and certain plants, but until 1997 it had never been shown for mammals. The creators of Dolly had taken an unfertilized egg cell with the nucleus removed, and fused that with the cell from the udder. The fused cell was made to divide and developed into a normal embryo. This was implanted into a surrogate mother, and it developed into a healthy lamb. This was the first time that genetic information from a fully differentiated, vegetative mammalian cell was used to give rise to a new, fully differentiated organism.

Building upon the "success" of Dolly, the next step came the same year from the same group in Scotland. The single, diploid cell originating from the adult sheep now was genetically altered before fusing with a denucleated egg cell. The fused cell was made to divide and to develop into an embryo, which was implanted into a surrogate mother. The resulting lamb, Polly, contains the human gene in every cell of her body.

The method resulting in Polly is seen as a major improvement as compared to the technology of the early nineties that led to the first transgenic bovine creature, Herman, carrying the human lactoferrin gene (lactoferrin is an important component in breast milk). At that time, genes were injected into newly fertilized eggs, and only in rather infrequent cases did the gene stably integrate and lead to a transformed animal.

Now that Dolly is an adult, questions are being raised regarding her health. In most respects she is a normal sheep. She is fertile and has given birth to a number of lambs. However, at a young age she developed arthritis, which usually does not occur in sheep until much later. Whether or not this is a consequence of cloning is as yet unknown. Her telomers are a little shorter than usual for an animal of her age, but whether this has an impact on her health and longevity is unclear.

Dolly in most respects was a normal sheep. She was fertile and has given birth to a number of lambs. However, at a young age she developed arthritis, which usually does not occur in sheep until much later, and she died at 6 years of age due to a progressive lung disease. Whether or not this is a consequence of cloning (which may not set back the biological clock) is as yet unknown. Her telomers were shorter than usual for an animal of her age, but whether this had an impact on her health and longevity is still unclear.

Many more animal clones have been generated in the mean time. Cloned cows appeared in 1999 and now there are cloned pigs that have been modified to reduce transplant rejection of pig organs in humans. Cloned pets (cats and dogs) have been created too. There are even cloned mules. The success with mammalian cloning has led to a large flood of responses, most of which are related to ethical issues regarding mammalian cloning. Indeed, what can be carried out with most mammals can also be carried out with humans, and therefore some very valid ethical issues on where to draw the line of the acceptable can be raised. An additional issue that has received a lot of attention is the use of human embryonic stem cell lines for medical purposes.

FISH FARMING

In some countries transgenic fish has been developed. This is quite easy, as there is usually no problem to get female gametes: just squeeze the female; no surgery, no microscopes. The egg cells can be just electroporated to introduce the desired

DNA. However, one does not need "high tech" approaches to increase aquaculture yields. Often it is sufficient to have fish male and female hormones produced in bacteria, and utilize these hormones. For some fish, after hatching a male fingerling exposed to estrogen will become female in appearance, while remaining genetically male. The "pseudofemale" can lay eggs producing viable offspring, in spite of the male chromosomes. Female fingerlings can be sex-reversed in the same way through exposure to testosterone, becoming reproductively viable pseudomales.

This application is attractive to fish farmers, who often want all-male groups of fingerlings: they grow faster than females and single-sex ponds mean that a second generation of fingerlings ("recruits") is not produced when the stocked fish becomes sexually mature. Recruits will eat, but will not be marketable by harvest time, thus bringing down production.

However, understandably consumers are not very enthused about having fish that has been exposed to hormones. To avoid having to hormone-treat fingerlings that will be harvested later, one can use a male parent with 2Y chromosomes and no X. All progeny will be XY and male, without needing any hormone treatment. YY males are indistinguishable from XY males and can be obtained from a normal male (XY) and a pseudofemale (a male that was treated with estrogen). Half of the resulting progeny is normal XY male, 1/4 is female (XX) and 1/4 is YY ("supermale"). XYs and YYs can be distinguished from each other by DNA typing. If one wants to produce progeny that is 100percent female, one can simply cross a "pseudomale" (XX) with a normal female.

T

U

V

W

X

Z